OILDORADO

Oildorado

Boom Times on the West Side

by William Rintoul

Valley
Publishers

OILDORADO

ISBN: 0-934136-07-6
(previously published in cloth under ISBN: 0-913548-49-9)

Valley Publishers
Division of Western Tanager Press
1111 Pacific Ave.
Santa Cruz, California 95060

Printed in the United States of America

For
My Wife
Frankie Jo

Table of Contents

Acknowledgments

I would like to thank Jack Rider of *Pacific Oil World* for permission to reprint material which previously appeared in that publication or its predecessor, *California Oil World*.

Thanks are due also to the reference staff of Beale Memorial Library, Bakersfield, for help in making research materials available to me, particularly Nina Caspari, who is in charge of the library's Historical Collection, and Mary Haas, who is in charge of the Geological, Mining and Petroleum Collection. I would also like to thank Richard C. Bailey, director of the Kern County Museum, Bakersfield, and his staff for help in furnishing photographs. Many others loaned photographs; their names appear in photograph captions. I would particularly like to thank Dennis McCall, Larry Peahl and Phil Witte for help with photographs and Mrs. Daisey Brown, Mrs. Margaret Mulroy, Mrs. O. C. L. Witte, Boyd Alexander, Claude Enyart, Bob Glover and Robert Gunning, among others, for sharing their recollections. I also want to acknowledge a debt of gratitude to the late Ann Bass McDonald, whose recollections of life in Reward form the basis for one of the chapters of this book.

William Rintoul
Bakersfield, California
December 1977

The Washington's Birthday Road Race

On a cold morning lightly laced with fog early in January of 1912, Kern County Deputy Sheriff Jack Basye, acting as chauffeur, set out from Bakersfield in the county Locomobile carrying a distinguished group of passengers on a trip that would take them through the booming West Side oil fields. The passengers included three county supervisors and the county engineer, who carried a pencil and pad with which to take notes.

The trip would be a long one, taking Deputy Basye, Supervisors Hart, Brite and Bush, and Engineer Evans more than one hundred miles over dirt and oiled roads from Bakersfield through the West Side oil towns of Maricopa, Taft, Fellows and McKittrick, and back to the starting point by way of Lokern and Buttonwillow. Though the trip would be slow of necessity because of Engineer Evans' note-taking, its reason for being was a much speedier trip that would be made over the same route in a matter of only a few more weeks. The purpose of the reconnaissance was to inspect the route in preparation for the big automobile race that was to be run over the same course on the coming Washington's Birthday. The race, only recently announced by the newly formed Kern County Automobile Racing Association, had already attracted statewide attention with the promise of being one of the most exciting automobile races ever run in California.

The trip took the party from Bakersfield past the east edge of Buena Vista Lake, where hunters complained there were millions of geese but only a sparse number of ducks, and those mostly teal. From there the group crossed the dirt flats east of Maricopa, where the ground was hard-packed in the absence of rain, the winter being one of the driest in years. Whatever the dry winter meant to farmers, it was welcome news to the race-conscious motorists, for the dirt roads posed enough problems without the added risk of running the race over muddy tracks.

A few miles east of Maricopa at the crossing over dry Santiago Creek, which normally carried water only after cloudbursts in the

Coast Range mountains that formed the southern boundary of California's great central valley, Engineer Evans noted that the existing bridge across the creek was too narrow. It was agreed that the bridge should be widened eighteen feet. The rest of the road was judged reasonably good, though in need of some grading to smooth out bumps.

The party proceeded into Maricopa, where the day before Marshal Carroll and his deputy, Thad Cheney, had raided Quong Sing Lum's Chinese lottery, declaring that this evil must go because it was pulling in the loose money of oilworkers. The raid, which eventually resulted in the levying of a $450 fine, overshadowed the arrival of a new shipment of books at the town library, including such titles as *Molly Make Believe* (Abbott), *Adventures of Sherlock Holmes* (Doyle), *Miss Gibbie Gault* (Boslos), *Sea Wolf* (London), *Buried Alive* (Bennett), *Sailors Knots* (Jacobs) and twenty-five volumes on socialism. The latter was a donation from the Maricopa Socialist Club.

The touring group from Bakersfield followed the planned racing route past the Carter Hotel in Maricopa to the First National Bank, turning there, as the racers a few weeks hence would do, to take the main road out of town toward Taft, seven miles away. A mile or two out of Maricopa the route took them past the sand-bagged crater and large holding sumps, some still containing oil, that marked the site of the Lakeview gusher. Some three months before the famed gusher had sanded up, ending an 18-month rampage during which the well produced an estimated nine million barrels of oil.

The scene that Deputy Basye and his passengers saw at the Lakeview gusher was vastly different from that of a few months earlier, when the well spouted oil, destroying the derrick and threatening the countryside with its uncontrolled flow. Now workmen labored to get the well back on production, probing the sand-bagged crater for the top of the six-inch casing that had been in the hole when it blew in. They had found the other casing strings, but the six-inch still eluded them. Speculation was that it had been ground to bits by the terrible friction of the uncontrolled flow and flung out with oil and sand during the eruption. Workmen probed nervously, for they were mindful of the gusher's force and none wanted to be in the way if the well got away again.

Though the scene of oil blowing high over the derrick was only a memory, it was soon to be duplicated on a huge silver trophy, ap-

The Lakeview gusher in all its glory was the inspiration for the scene engraved on the Lakeview trophy to be awarded along with $1,000 in cash to the winner of the Washington's Birthday road race through the West Side oil fields. (Photo from Kern County Museum)

To workmen who paddled across holding sump filled with Lakeview oil,
the West Side, circa 1912, seemed afloat in a sea of oil. (Photo from
Clarence Williams Collection)

propriately named the Lakeview trophy, that would be awarded to
the first place winner in the coming automobile race. The trophy,
valued at $500, was even now being prepared. On one side there
would be engraved a depiction of the Lakeview gusher in all its glory;
on the other, the names of the trophy's donors, J. Fried, Parker
Barrett, A. E. Hodgkinson and J. M. Dunn, who were among the
original backers of the great gusher.

The inspection party from Bakersfield followed the detour around
the gusher site, making a sharp turn to cross the Sunset railroad
tracks, following the tracks a short distance, then making another
sharp turn to recross the tracks and pick up the regular Maricopa-
Taft highway. Evans ventured that the sharp turns and grade cros-
sings would pose even worse hazards than the normal road conditions and suggested that the original road past the gusher, closed
during the great flow of oil, be reopened in time for the race. The
supervisors took the recommendation under consideration and noted, as did Evans, that the Maricopa-to-Taft stretch desperately
needed resurfacing. There were chuckholes a foot deep, and oiled
ridges high enough to threaten the undercarriage of a car.

In Taft, which had been incorporated hardly more than a year
before with a population of 750, the party from Bakersfield found
excitement running higher than might have been expected over the
coming race. The truth was, as of a few days before, there was
suddenly less than usual in Taft to get excited about. As of the New
Year, Constable Sam Ferguson, at the behest of the town's more
orderly element, had closed down the dance halls, sending what the
Daily Midway Driller, the town's newspaper, called the "macque-
raux" and their women on their way. The term was a corruption of
the French word, "maquereau," which meant "mackeral." Though
the dance halls were gone, saloons still lined Main Street, down which
racers would roar, but they did not completely fill the gap. In fact,
there even seemed to be an element of righteousness creeping into
the saloons. One advertised itself as "the only strictly family liquor
store on the West Side" and offered to make free deliveries anywhere
in Taft.

On the part of the town's merchants, interest in the coming race
was tempered with a more serious consideration, namely, a rate
case in which they were joined by Bakersfield merchants against the
Southern Pacific and Santa Fe railroads, co-owners of the Sunset
railroad. The two companies had extended their spur line three

Long after the derrick toppled, Lakeview No. 1 continued to gush oil,
flowing an estimated nine million barrels in eighteen months. (Photo
from Clarence Williams Collection)

years before to serve oil development in the Midway district some seven miles beyond Maricopa. They had named the siding at the terminus of the extended line "Siding Two." The false-front saloons, frame buildings and tents that clustered around the siding had coalesced into the city of Taft, named in honor of the incumbent president of the United States.

Taft depended on the railroad not only for its access to the outside world, 78,669 passengers having traveled over the spur line during 1911, but also for most of its oil well supplies and merchandise as well as its drinking water, normally comprising seven tank cars a day. The merchants charged the line with making an excessive profit, citing a return of 54 percent on the investment the last fiscal year. They found it hard to understand why the freight rate from Los Angeles to Pentland siding near Maricopa, a distance of more than 150 miles, was 54¢ and only a penny cheaper, 53¢, from Bakersfield to the same siding, a distance of 40 miles; or why it was only one penny more from San Francisco to McKittrick, a distance of about 300 miles, than from Bakersfield to McKittrick, a distance of 35 miles. They were taking their case to the State Railway Commission, which was convening in Bakersfield, and they had lined up various witnesses in their behalf, among them Charles Fox, the editor of *California Oil World*, whose testimony on the bright future of the West Side oil fields was expected to bolster the case for a more equitable shipping rate.

The West Side merchants, along with their outrage over rates, had another complaint. The railroad, almost as if it meant to add insult to injury, had on the first of the month changed the schedule, bringing the late train into the West Side at 8:15 at night. Merchants complained this ruined mail service. Before, it had been possible to post a letter in Taft in the morning which would be delivered in Bakersfield before lunch, and the recipient in Bakersfield, if he took pen in hand without delay, could post a reply in time to catch the afternoon train, the letter being delivered in Taft that same afternoon. With the change in train schedule, it was no longer possible to have same-day mail service, and the merchants were outraged.

Beyond Taft the Bakersfield party headed for Fellows, five miles away, traveling through the Midway oil field, which was the biggest producer in the state, having put out almost 21 million barrels in the year just ended, or substantially more than any other field and almost four times as much as the adjoining Sunset field. After mo-

Taft depended on the Sunset railroad not only for access to the outside world but also for most of its oil well supplies and merchandise as well as drinking water, normally comprising seven tank cars a day. (Photo from Phil Witte)

toring through Fellows, the group passed close by the Buick No. 2 gusher on Sec. 32, 31S-23E, which had blown in six days before from 3,319 feet and was making an estimated 10,000 barrels a day, throwing oil 200 feet over the top of the derrick with a roar that could be heard for miles around. Men were building sumps to contain the oil, and the touring party decided the well posed no immediate threat to the race course.

The party proceeed to McKittrick, ten miles beyond Fellows, and from there eight miles on to Lokern, with Evans noting that the stretch just beyond McKittrick needed much leveling. By the time the party reached Bakersfield, tired and dusty, they had traveled 106 miles, half the distance the racers would travel going around the course twice. The supervisors and county engineer, pleased with the enthusiastic reception they had been given in the West Side oil fields, were confident the coming event would be a success. There was at least one side effect. Deputy Basye, the driver, had been caught up with enthusiasm and on the conclusion of the trip, he announced to his passengers that he himself had decided to drive in the race, provided he could find a sponsor.

The deputy's decision to enter caused no particular excitement, but the same could not be said of a news report that was published soon after by a Los Angeles newspaper and widely circulated in the Kern County oil fields.

The newspaper reported that Bert Dingley, the well-known Los Angeles racer, had sold his famed Pope-Hartford racing car to a party identified only as a Bakersfield autoist. The Pope-Hartford was the same racer in which scarcely more than a month before Dingley had set a world's record for a one-mile run on a dirt track. He had chalked up the record on the Phoenix, Arizona track, traveling the mile in 52¾ seconds. There was speculation that Dingley might have sold the car to Tom Klipstein, one of the racing enthusiasts in Bakersfield, and the conclusion was that Klipstein planned to enter the racer in the coming run through the oil fields. The thought was enough to give pause to anyone contemplating entering the event, for the powerful Pope-Hartford, proven at Phoenix, would be a hard competitor to beat. To all inquiries whether he was the purchaser, Tom Klipstein responded with silence and a noncommittal smile, leading to speculation that he intended to put one over on his racing compatriots.

The rumored entry of Bert Dingley's Pope-Hartford might have overawed some would-be entrants, but it did not faze E. D. Burge, an oil operator and automobile enthusiast. Burge announced he would enter a National Roadster in the coming race. He said he himself would sponsor the car, for the makers of the racer had recently announced they were retiring from sponsoring cars in races. A National spokesman had stated, "National has won everything offered in the racing world and the car has nothing further to win by staying in the game." There was some truth to the statement for in the Los Angeles-to-Phoenix run that had preceded the same Phoenix race in which Dingley's Pope-Hartford had set its world's record, Harvey Herrick had driven Burge's National to a smashing triumph, making the 524-mile run in the shattering time of 21 hours and 15½ minutes.

In the brash West Side oil fields, the prospect of competing with Dingley's Pope-Hartford and Burge's big National did not disconcert men who almost overnight had turned the local fields into the biggest in the state. The West Siders could not be accused of being backward. Tommy McGinn was busily forming an association to open offices in London, England, for the purpose of advising investors on matters pertaining to the West Side fields, and Al Thackery, who ran the Green Store on Main Street, which advertised "no dry

The condition of the oil field roads posed special hazards for the fifteen racers entered in the grueling Washington's Birthday run. Some thought lighter cars like the Buick would outlast the faster National and Pope-Hartford. (Photo from Kern County Museum)

goods," was off to San Francisco with what the *Driller* described as a "bankroll as big as your arm" to bring the Jack Johnson-Jim Flynn heavyweight championship bout to Taft. The West Side fields were attracting an increasing number of visitors, including some who came to see how West Siders did it, like Kazutaka Ito and Teisuke Watanabe, Nippon Oil Company officials who had come on a tour of inspection accompanied by Dr. E. A. Starke, Standard Oil geologist. Others came to solicit West Side know-how for their own purposes, among them a farmer from Glenn County in Northern California's Sacramento Valley who swore there was gas under his land and wanted to interest a driller.

It so happened there was at least one Pope-Hartford in the oil fields, where men prided themselves on the cars in which they sped through the fields, even as their fathers had prided themselves on their horses. The owner of the 40-horsepower automobile was Jack Maddux of the Dover & Wilson garage in Taft. He was no newcomer to the racing game, having driven all the courses at San Francisco, San Rafael and San Jose. As if that were not enough, he had driven for two years in Alaska, which he claimed gave him experience to buck anything he might encounter in the coming race. Maddux announced his entry, and began training runs without delay.

There was talk that big cars might be good where roads were in passable condition, but that the run through the oil fields would prove too much for them, and that the kind of car it would take to win would be a smaller car with the ability to take it when the going got rough. The car that came to mind was the Buick. A month before, a driver for American Well & Prospecting Company had had the misfortune to go over a 250-foot grade at Cat Canyon near Santa Maria in the company's Buick. The car had been badly damaged, the driver almost killed. They had taken the driver to the hospital, where he was still recovering. They had hauled the battered Buick to the Taft Vulcanizing Works, where Manager D. R. McClary had taken a long look at twisted axles, smashed wheels and broken lugs, and gone to work. The car was now back on the road. If one wanted a tough car, Buick was it.

W. S. Lierly, who was the owner of a new Buick 30, was prevailed upon in the name of community pride to enter his automobile in the coming race. The car was promptly dispatched to Ramsey's garage, newly purchased by McClary, to be stripped for racing. The driver would be Dick Gochenouer, who had retired three years before after a successful career as a racer during which, among other feats, he had established a record for the Tucson-to-Phoenix run that still stood. Gochenouer announced he would come out of retirement to drive the Buick, and the job of stripping down the speedy car continued apace. Meanwhile inquiries about the race arrived at the Bakersfield office of the Kern County Racing Association from as far away as Cincinnati, where the company that manufactured the Cino inquired about entering a model. A Nyberg 40 became the first official entry, sponsored by McKee & Scott, agents for the Nyberg in Bakersfield and Fresno.

One of the few portions of the race course that had been surfaced was the seven-mile stretch between Maricopa and Taft, which had been heavily oiled, perhaps because of the availability of oil from the surrounding wells. The stretch was, coincidentally, the worst of the entire 106-mile course. As the supervisors had noted on their inspection drive, traffic and the sun had caused the oiled surface to become potted and ridged, so that it was not uncommon for a car's wheels to sink so deeply in ruts that the undercarriage dragged.

The county's roadmaker, J. W. Greene, acting on orders from the supervisors, concentrated his efforts on the seven-mile stretch, assigning eleven men and six teams to the job of setting the road in order. For the exacting job of cutting down the ridges, Greene rigged a device that would prevent the grader's blade from cutting off more than an inch or two at a time, which he proudly demonstrated to all who would watch. Greene's men also reopened the road past the Lakeview gusher, eliminating the sharp turns over the Sunset railroad tracks. While work continued on the Maricopa-to-Taft stretch, other county crews put in long hours on other portions of the course, hurrying to get the highway in shape for the race.

In Taft, concern was expressed over deep pits in Main Street, where the town's saloons clustered and along which the racers would speed. It was suggested that the Santa Fe, which presently operated the Sunset railroad, might be called into service. A huge mound of dirt by the Third Street crossing might be labeled unsightly, and the railroad forced to dispose of the dirt by spreading it over the deep depression in the road at Fifth and Main. The possibility of getting the railroad company to move quickly enough to get the job done before the coming race appeared doubtful.

The town recorder, Sam Birehard, came up with another suggestion, stating that he would see the road in shape if he had to put

The Big National, shown here in an earlier race, was the favorite to win the 212-mile run through the oil fields. The machine was fresh from a win in the Los Angeles-to-Phoenix road race. (Photo from Kern County Museum)

Planners warned others to keep off the road on the day of the race, hoping racers would not encounter teams like O. P. Goode's, shown moving a bunkhouse from Midoil to the CCMO camp near Fellows in 1912. (Photo from Phil Witte)

twenty men on a chain gang to move dirt from the turn at Oil Well Supply Company's headquarters near the Sunset crossing. The chain gang did not presently exist, but the implication was that it would not be hard to form one from men who spent their time in saloons on Main Street. The threat against saloon habitues' freedom was only one threat the drinking group faced. The Ladies Improvement Club was in the process of circulating petitions demanding that the town's saloons be confined to Main Street and not allowed to spread to adjoining Center Street. Mrs. L. P. Guiberson led the signature gatherers. This was not the only move by the ladies' club aimed at bringing a higher cultural level to Taft. The ladies also were preparing to present the play, "The Matrimonial Club," which would feature authentic costumes dating back to the 1840s.

Getting the roads in order was only part of preparation for the big race. There were other details like selecting officials and getting the race sanctioned by the American Automobile Association. The choice for race referee was Ed Kuster, attorney for the Automobile

Club of Southern California. Kuster accepted the assignment, arranging to be on hand from Los Angeles a day or two before the event to inspect the course and meet with drivers and mechanics to make sure all understood the rules. There were permissions to be secured, including one from the city of Bakersfield to close H Street from 19th and H in downtown Bakersfield, where racers would start and finish, for the street's entire length out of the city, and another from the county for the closing of the entire race course, which was secured for the time from eight in the morning when the first racer would be flagged on his way until four in the afternoon, when the last racer presumably would have crossed the finish line.

There were arrangements to be made, too, for the guarding of the race course so spectators would not get in the way. National Guardsmen, to be armed with red and white signal flags, were given the assignment on county roads. Police drew the duty in Bakersfield. On the West Side, the West Side Motorcycle Club appointed J. C. Snook, L. C. Jewett and J. W. Woolridge to set up patrols by club members. The motorcyclists also announced plans to hold a dance after the race at the Maricopa opera house, featuring the music of L. C. Sobek and his five-piece orchestra.

None of the details posed any particular problem except one. This concerned prizes to be awarded winners. H. C. Katze of the Kern County Automobile Racing Association had announced there would be three monetary prizes. The winner, in addition to the Lakeview trophy, would receive $1,000; there would be a $700 prize for second place, and a special $300 prize for the first light car to cross the finish line. As February 22 approached, consternation set in among sponsors of the race. Where was money going to come from for prizes? The entry fee was $25 per car, hardly enough, considering the dozen or so entries expected, to put together the prize for the light car, much less the $1,000 promised the winner, or the $700 for second place. The problem was that many of the merchants who might normally be expected to contribute were temporarily low after putting up money for the rate case against the railroad. Contributions were coming in, but they were small, $10 from this one, $5 from that one. With the race almost at hand, sponsors took to the road themselves, making frequent trips to the West Side to drum up all the support they could from local merchants.

Worried race sponsors were not the only ones on the road to the West Side oil fields. As the entry list swelled to a dozen, the county highways that would shortly form the race course were the scene of death-defying trial runs. Speed limits were, of course, suspended for those testing their machines, and the stream of goggled racers roaring through the countryside without interference from the law was in ironic contrast to the Oakland motorist who drew a $20 fine for speeding through Bakersfield at the pace of 28 miles per hour. The rush of racers through the towns on the West Side prompted the Maricopa board of trustees to seek relief. The trustees, fearful that someone would be hurt, notified racers they would be expected to reduce their speed during trial runs to ten miles an hour through Maricopa or face the legal consequences.

There were rumors of record times. Jack Maddux, driving the Pope-Hartford 40, reportedly chalked up a time of only 45 minutes for the 39-mile stretch between McKittrick and Bakersfield, or almost a mile a minute. Maddux, with H. Dover as mechanic, arrived back in Taft covered with sticky black oil and dust, triumphantly clicking a stopwatch as he crossed the point where he had started. Dover said he had never gone faster. The appearance of the car bore out that it had been going at high speed.

It was reported that Phil Klipstein with Bobby Howland as mechanic had covered the 43.7-mile stretch between Bakersfield and Maricopa in 57 minutes. Klipstein was driving a Mitchell. He credited his success in the gray-and-black car to the use of Kelly-Mitchell tire casings, which, he said, proved their worth on the "heavy" roads.

Dick Gochenouer of McClary Garage went out with the Buick 30, which, though it had 10 horsepower less than the Pope-Hartford, was considered a worthy competitor. Though Gochenouer disclaimed any record times, it was reported he had made the run in speedy style. J. D. Lines rode with Gochenouer as mechanic. He was an experienced repairman. The Buick, owned by W. S. Lierly and completely overhauled and stripped for racing, was said to be in top shape.

Hardly an hour passed that a racer could not be seen on the road. There was Tom Marsh, driving a Reo 30, who roared into Taft with Harry Croson and Earl Jackson along as what the *Driller* described as "mechanics and ballast." The Reo had its backers. The machine recently had set a world's record for a 50-mile run over a dirt track making the distance on the San Diego track in 56 minutes 12-4/5 seconds. There were others on the road too. Gale Hammett in McKee & Scott's Nyberg. C. Klein in C. H. Kaar's Flanders. T. Lyman in

The quiet that prevailed on this Maricopa street early in 1912 would soon be broken by the roar of racing cars competing for the Lakeview trophy in a 212-mile run through the oil fields. (Photo from Kern County Museum)

Ben Brundage's Ford. W. M. Moore in James Arp's Knox. "New York" George Blast in Webster Garage's Buick. A. I. Robb in a Buick. C. Hollenbeck in a Buick.

The greatest excitement occurred when the big National that had won the Los Angeles-to-Phoenix race appeared on the scene, driven by Harvey Herrick, who, though he himself would not drive in the Washington's Birthday race, was prepared to train another driver, Glen Packer. Oilman Burge's machine, stripped down and ready to race, was immediately established as the favorite. The big National made the circuit, attracting no paucity of attention as it took corners through Taft at high speed. It appeared that whatever chance any other racers had was rapidly narrowing.

The coming race was a hot topic of conversation in the West Side oil fields, sharing interest with such things as the experiment that Santa Fe was conducting with a Westinghouse electric motor to furnish power for drilling at a rig on Sec. 36, 31S-22E, the "juice" having been used to make a phenomenal 450 feet of hole in 73 hours; or the $1.5 million pipeline that was to be laid to carry gas produced by Honolulu Oil to Los Angeles; or the disappointment at the Buick No. 2 gusher, where flow had turned to valueless emulsion.

Even as enthusiasm mounted for the Washington's Birthday road race, the electrifying news came that there might be yet another entry from the West Side. Scott Weaver, a former driver for Esperanza Oil Company and known as one of the fastest drivers in the oil fields, was talking of entering his 50-horsepower Kline. He had the car in the garage and George Barr was overhauling it. The Kline 50-horsepower engine had a 10-horsepower margin over Jack Maddux and the Pope-Hartford and a 20-horsepower edge over Lierly's Buick; the car seemed to have the best chance to beat the big National. Could Barr complete the overhaul in time?

Three days before the race, the *Bakersfield Californian* reported, "When the big Kline car with its six cylinders shooting fire sped out of the Taft Vulcanizing Works, interest took another jump. The Kline, with the power of 50 horses, is a promising car and from its strong appearance is winning popularity." The car had 36-inch wheels in front, 37-inch behind, giving it something of the appearance of a cannon lying low between huge wheels. Scott Weaver would drive; Roy Herndon would be the mechanic.

While West Side hopes soared with the roar of the Kline on the roadway, the hopes of the race's sponsors soared with an idea that might save the embarrassment of having to decrease prize money. They announced they would sponsor an air show to raise prize money and promptly lined up two aviators from Fresno to put on the show on the weekend before the big race.

The two-day show at Edison, an agricultural district several miles east of Bakersfield, would feature races pitting an airplane against a car and a motorcycle, acrobatics, and in a more serious vein, a demonstration by Aviator Frank Bryant of the airplane's practical use as a weapon of war. The plan called for the outline of a battleship to be marked out in lime on the ground, drawn to the actual specifications of the ship. The aviator would bomb the ship from high altitude, showing how it could be done. The "bombs" would be oranges grown in the Edison district. Special trains would carry spectators to the aviation meet, with the $1 fare covering both train fare and entry to watch the air show.

To help promote the show, sponsors drove Aviator Bryant out to the West Side oil fields, where he was quoted as saying, "This West Side section is the best country in the world. I was in Alaska during the first big gold rush and I have lived in border country for many years, but the West Side beats them all. I like the spirit of enthusiasm and boost that marks the strides of progress. It is a wonderful field and if I wasn't busy running aeroplanes, I surely would settle in this section and remain here the rest of my life."

Two days later, Bryant probably wished he had given up flying in favor of the oil fields. While circling the crowd of spectators at the air show in his Curtis Biplane, he lost power and had to make a deadstick landing in a plowed field a mile from the viewing stand. Though he walked away unhurt, the plane was badly damaged, and the air show proceeded at a reduced pace with only one plane.

When sponsors added up receipts, there was bad news. They had taken in $991, of which $500 went to the aviators and most of the rest for payment of other expenses incurred in staging the air show. In the nick of time, West Side merchants and oilmen, whipped to fever pitch by the racers roaring through their communities, dug into their pockets and came up with the money, insuring the race would go on as planned. The same could not be said, however, for postal service out of Bakersfield. It was announced there would be no deliveries along the rural route on Rosedale Highway the day of the race, the reason being that the highway was to serve as part of the race circuit.

In Bakersfield, two more entries were noted, swelling the field to fifteen. The machines were Stutzes, including one by W. W. Hatton, another by Tom Klipstein. G. E. Ruckstell would drive the Hatton Stutz. Klipstein named Deputy Sheriff Jack Basye as his driver. The entry of the Stutzes caused some speculation, for the Stutz was the newest of the cars in the field. The very first Stutz had made its debut less than a year before in the inaugural run of the Indianapolis 500. At Indianapolis, the Stutz had finished out of the money in 11th place, but the four-cylinder car had survived the race without a single mechanical adjustment and there were those who said it would have been a winner if it had not been for ten pit stops to replace tires.

Betting favored the big National, with so few willing to bet against the car's winning top honors that most of the brisk action involved the finishing order of the other machines. Jimmy Murray had $40 that said neither Stutz would win, and Tom Klipstein covered the bet. F. C. Venator of Bakersfield had $250 which he was willing to bet against $200 that the Pope-Hartford would not finish first, second or third, but could not get a taker.

The day of the race dawned cool and clear with no hint of rain or fog, an ideal day for racing. In the early morning hours, Scott Weaver and his mechanic, Roy Herndon, set out from Taft in the powerful Kline, carrying the hopes of West Siders with them and their formidable racing machine. Weaver opened up the car, roaring over the road on which the racers would soon run. Five miles out of Bakersfield, the road made a sharp turn to the right, crossing a bridge over a ditch. Weaver, traveling at seventy miles an hour, missed the turn, plunging into the ditch. He was thrown clear, breaking an arm and two ribs. Herndon was uninjured. The big racing car was badly damaged. The vehicle, which had been scheduled as Number Thirteen in the day's race, was scratched from the entry list, leaving a field of fourteen cars.

In Bakersfield, half an hour before the race was to begin, Jack Moore, foreman of the Webster Garage, was driving the Buick that was the garage's entry, warming it up for "New York" George Blast, who was to drive in the race. While traveling at a high rate of speed, Moore plowed into the cable that had been put up to protect spectators at the corner of 20th and H Streets. He was thrown clear, breaking three ribs. The car's steering mechanism was damaged. Mechanics feverishly went to work to repair it.

Meanwhile a crowd later estimated at 8,000 persons began gathering behind cables along H Street, waiting for the race to begin. Along the course, National Guardsmen took up places, flags at the ready so they could signal when cars were sighted. On the West Side, the motorcycle club took up its positions. Virtually everyone took the day off, with business in the oil field communities coming to a halt, except along Main Street in Taft where saloons offered a front row seat with refreshments near at hand.

A few minutes before eight o'clock the Nyberg came to the line at 19th and H Streets, and the crowd cheered. Promptly at eight the car was off. Scarcely had the orange-tasseled caps of Driver Hammett and Mechanic Kellam dropped out of sight down the road in a cloud of dust before the No. 3 car, taking the place of the Webster Buick, over which mechanics still worked, rolled to the line. It was the big National. The car was quickly off, its bright blue radiator snapping in the sun whenever the clouds of smoke rolled away sufficiently to see it. No. 4 was the Flanders; it was flagged off five minutes later and had traveled only five blocks when it was stopped by a freight train on the Santa Fe tracks. The car lost more than a minute while it waited for the train to pass. No. 5, the Robb Buick,

For Taft, the Washington's Birthday road race offered the greatest excitement in the town's brief history. Fifteen racers would roar down Main Street making a circuit run from Bakersfield through the West Side oil community and back again. (Photo from Kern County Museum)

got away in a cloud of dust and was up to 45 miles an hour within the first two blocks. At five-minute intervals, others followed, including Ford No. 6, Buick No. 7 with Dick Gochenouer driving, Mitchell No. 8, Reo No. 9, and Stutz No. 10 with Jack Basye driving. Knox No. 11 would not start and was scratched, leaving thirteen cars, and one of those—the Webster Buick—doubtful. The Hollenbeck Buick No. 12 was next off, followed by the Pope-Hartford, filling in as No. 13 for the demolished Kline. No. 14 was the Hatton Stutz driven by G. E. Ruckstell. The last to start was the Webster Buick. It was the thirteenth and final car to begin the race.

In Maricopa, it seemed as if the entire town and all the population of surrounding oil leases lined the race course. A cheer went up when the first plume of dust was sighted on the flats east of town. It was the Nyberg, thundering into town. With a rumble and a roar, the speeding car, with its dust-begrimed driver, Hammett, sitting grimly at the wheel, came abreast of the crowd. Mechanic Kellam waved in response to the shouting spectators, and the machine shot by.

The next car through was the National, thundering by with throttle jerked wide open, covering the 43.7-mile stretch in 53 minutes. The crowd greeted its appearance with cheers and shouts.

The next car to reach Maricopa was the Nyberg, traveling in the wrong direction. A mile or two out of town while racing toward Taft, the machine had broken a rear spring and its gas tank. Hammett limped back to the Maricopa garage, where it was quickly apparent the car was out of contention, leaving twelve cars in the running.

Next into Maricopa, traveling at almost a mile-a-minute, was Stutz No. 10, which had covered the distance in 45 minutes, or eight minutes faster than the favored National. Cheering crowds greeted the speeding car as Jack Basye thundered through town.

Next came the little Flanders, shooting through town with a shade better time than the National, followed by Buick No. 5, Buick No. 7 and the Pope-Hartford. Buick No. 12 roared into Maricopa spewing oil. It was discovered the car's oil tank was punctured. The car was through for the day. The race now had eleven cars.

It was soon apparent that the field was even smaller. The Ford staggered in, barely making it on two cylinders. It was obvious the machine was out of contention, leaving the race to a field of ten cars.

Beyond Taft, the race course passed through Fellows, where the prospect of the Washington's Birthday road race stirred keen interest. (Photo from Phil Witte)

Spectators at Maricopa waited in vain for Buick No. 16, which had been the last car to leave Bakersfield because of the accident before starting time. East of Maricopa, "New York" George Blast missed a turn and went into the ditch doing sixty miles an hour. Both he and his mechanic, William Gray, were injured. W. B. Moore rushed them to the Maricopa hospital, where it was found that Gray had suffered a broken shoulder, Blast severe lacerations. The wreck was the second and final one of the day for the Buick. The race now had nine competitors.

In Taft, a large crowd enthusiastically greeted the arrival of the first racer, the National. The National was followed by Stutz No. 10 and the Pope-Hartford. The *Driller* reported, "Never before in the history of Taft was the mad enthusiasm shown that was exhibited when the big cars came plunging through the town. The cheers and ringing of bells and blowing of whistles were deafening." Betting was brisk, with the Stutz entries beginning to draw support.

In Bakersfield a large crowd gathered in front of the offices where the *Californian* was published. The newspaper had made arrangements to receive reports by telephone from observers in Maricopa, Taft and McKittrick, and each report of the arrival of a racer, together with the time, was quickly posted. A shock of disappointment went through the crowd when the report was posted that the gallant Flanders was through at McKittrick. Driver Klein was running second when he went out with a punctured gas tank. There were now eight cars in the running, and the race was still in its first lap.

If the Bakersfield crowd took the news hard about the Flanders, their gloom was small compared with the disappointment that was spreading through the West Side oil fields. Word came by telephone that Jack Maddux, who had gone through McKittrick driving like a madman, was in the ditch three miles from Lokern. Though Maddux was not seriously injured, the Pope-Hartford was finished. There were now seven cars left, and of these, only the two Stutzes and the National appeared to have a chance for the big money. C. J. Jacobs of Maricopa offered $500 to $50 that a Stutz would take the top prize. He had no takers.

In Bakersfield, the huge crowd that had witnessed the start waited for the first racer to complete the first of the scheduled two laps. Some people drifted off to the dedication ceremonies that were being held for the new $400,000 county courthouse, but most chose to wait for the return of the racers. Shortly before 11 o'clock, less than three hours after the Nyberg had roared out, the red and white signal flags fluttered down Rosedale Highway. A streak of dust appeared. The big National roared into Bakersfield, completing the first lap in 2 hours, 37 minutes, 30 seconds. The car made a quick pit stop. An inspection revealed that the front spring was broken. Everything else seemed in order, and Packer roared off. Buick No. 5, which had been the fourth car to leave, was next across the starting point. Mechanic Ellis reported they had passed Stutz No. 10 in a ditch at Rio Bravo. The Buick made a quick pit stop at which it was necessary to adjust a bent shaft. It was speedy work, but in the haste a helper spilled most of a can of gasoline, causing some delay.

Ten minutes later Stutz No. 10, seemingly none the worse for the ditching at Rio Bravo, roared into town, delayed but a minute and was off. Not long after, Stutz No. 14 rolled in, made a brief pit stop and resumed the run.

The race appeared to be between the National and the Stutzes for the top two spots, with Buick No. 5, Buick No. 7, the Mitchell and the Reo in the running for the $300 prize for the first of the light cars to finish.

From Maricopa came a telephoned bulletin. The National had broken an axle short of town and was through for the day. There were now six cars in the running, with the Stutzes all but assured of top money. Another bulletin came in. Stutz No. 14 was out with a blown cylinder 18 miles into the second lap, leaving five cars. And then from McKittrick came the report that Stutz No. 10 was in trouble. The clutch appeared to be going out. Driver Basye was battling to stay in the race. In Maricopa, C. J. Jacobs undoubtedly thanked his lucky stars that no one had taken him up on his offer to bet $500 to $50 that a Stutz would take top prize.

At the finish line in Bakersfield there was talk that none of the cars would finish before 3 P.M. At about two o'clock, a shout went up from the crowd. Red and white signal flags fluttered. Dust rose in the distance, and spectators cleared the track. In a cloud of dust a car roared in. "Stutz, Stutz," shouted the crowd.

And then a disappointed sound arose. It was not Stutz No. 10 at all. For a moment hearts stopped among those who had bet on the Stutz. The arriving racer was the Nyberg, repaired and back on the road, doggedly finishing the first and only lap it would finish that day.

Washington's Birthday race course included the main street of McKittrick. (Photo from Kern County Museum)

The bettors, nerves jangled, settled back. At 2:30 a shout went up. Dust was rising. A car was coming, and coming fast. It was Stutz No. 10 with Jack Bayse at the wheel. The big car crossed the finish line a few minutes after 2:30 P.M., making the 212-mile course in the shattering time of 5 hours, 44 minutes, 59 seconds. The Stutz was the winner, and the only one of the big cars to stand up under the punishment of the gruelling race through the oil fields.

An hour and twenty minutes later the Mitchell with Phil Klipstein at the wheel crossed the finish line, winning both second prize of $700 and the special prize of $300 for the first light car to finish, thereby equalling the amount of money won by the Stutz. Buick No. 7 followed, finishing just out of the money. Driver Gochenouer reported he had lost an hour at Conners Lane 15 miles out on the second lap. A tire had blown, and both rings had come off. He had recovered the heavier ring, but the lighter one had been lost. He and Mechanic Lines had searched a half-hour before they had found it. They had no sooner replaced the tire and gotten a mile down the road when they had another blowout. This time the smaller ring had rolled into a farm pond. They had found it after a half-hour search, and Lines had waded in to retrieve it and get them back on the road.

At 4:00 P.M., the time at which the road was to be reopened for general usage, judges declared the Washington's Birthday road race over. Only three of thirteen cars had succeeded in finishing.

For the Stutz automobile, the race was a better advertisement than any that might have been purchased in the newspaper. The message was not lost on oil men. The day after the race, H. H. McClintock, who was prominent in the West Side fields, made a special trip to Los Angeles to purchase a new Stutz which he drove back that same evening. Soon afterward General Petroleum Corporation paid $16,000 to purchase six of the machines for its superintendents, explaining that the company wanted the best there was.

For Jack Basye, the man who had guided the powerful Stutz 212 punishing miles to victory, the aftermath of the race proved embarrassingly like the proverbial story of the man who survived a trip over Niagara Falls in a barrel only to slip on a banana peel and break his leg.

On the evening of the day after the race, Basye and his fellow deputy, Ed Grandy, who had ridden with him as mechanic, were hurrying down Chester Avenue in Bakersfield on their way to the county jail when, in the neighborhood of the Kern Valley Abstract office, Basye's revolver fell from his pocket to the cement sidewalk. The gun discharged, and the steel-capped bullet passed through the calf of Basye's left leg to lodge in the thigh of his right leg. The deputy hobbled to the jail, where he was given first aid before having the bullet removed by Dr. N. N. Brown. The accident left Basye to enjoy the accolades of his racing triumph from the confinement of a bed in San Joaquin Hospital.

When Hollywood Looked at Oil

Many a strapping youth sought fame and fortune in the West Side oil fields, but none enjoyed a success story any more heartwarming than that of a likeable young man who was known to residents of the booming oil town of Taft as young Arbuckle.

Arbuckle came to Taft in the days when gushers roared in out of control making thousands of barrels of oil a day, when drinking water sold for more than crude oil, and whiskey outsold them both in a score of well-patronized saloons. No one bothered to fix the date of the penniless Arbuckle's inauspicious arrival, not even Arbuckle, but it seems to have been not long after the town was incorporated, an event that occurred in 1910, the same year the Lakeview gusher blew in as the world's most spectacular gusher.

Soon after his arrival, Arbuckle fell hopelessly in love with a beautiful young lady named Pauline. She was petite, demure and thoroughly feminine. Pauline's father was a capitalist, a thoroughly wealthy oil man. When Arbuckle asked for Pauline's hand in marriage, her father, fearing the erstwhile suitor might be nothing more than a fortune-seeker, said, "Nay, nay. No man may have my daughter until he works himself from the ground up."

Arbuckle repaired to the oil fields, where he approached a man named Frank Whitney, who was superintendent for Standard Oil Company, and asked for a job. Whitney, obviously impressed by the young man's attitude, forthrightly put Arbuckle on as a driller. Considering Arbuckle's lack of experience, it was not too surprising that the well he was working on quickly got away, blowing a stream of oil high over the crown block. To complicate matters, the wild well caught fire. Many a lesser man might have quit his job in despair, but Arbuckle, undaunted, singlehandedly succeeded in capping the burning gusher. The feat made him a hero in the eyes of his fellow oilworkers. It is doubtful, however, that the heroic effort endeared Arbuckle to Whitney, for it was not long after that Arbuckle, following additional seasoning in the transportation and pipeline departments, replaced Whitney as Standard Oil's superintendent.

As superintendent, Arbuckle was immensely popular with the men. Whatever their hostility to other bosses, they were extremely fond of Arbuckle, so much so that they accorded him what was perhaps the greatest honor that could be bestowed on a man in the sports-conscious West Side oil fields. They made him captain of Standard Oil's baseball team. Arbuckle responded to the honor with as much verve as he had displayed in taming the burning gusher. He led the team to a pennant victory in a league that included such arch rivals as Maricopa, Santa Fe, Kern Trading & Oil and Associated.

To some, it might have seemed as if Arbuckle had more than proved himself. He had risen in the shortest possible time from driller to superintendent of the most powerful oil company in the West Side fields. He had demonstrated his courage beyond any doubt in a one-man assault on a burning well. He had convincingly displayed his athletic prowess on the ball diamond. What more could anyone ask?

As luck would have it, another challenge came along. About this time the proud city of Taft faced the task of electing a mayor. To the field of candidates, Arbuckle's name was added. Could there be any doubt about the outcome? Arbuckle proved himself an able orator, delivering a masterful campaign speech to enthusiastic production hands in the Section 14 yard. He established himself as a practical politician by making a visit to the grounds of Taft's new $50,000 Conley School where the principal, Jack Hamilton, obligingly turned out the children. As each child filed from the school door, Arbuckle graciously handed out a small present.

Now the town of Taft was a gambler's town, with odds quoted on everything from the weekly boxing matches staged by promoter Izzy Rehfeld to the exact moment when the latest gusher, wherever it happened to be, might be brought under control, but it is doubtful if there was any wagering on the mayor's race. Who would have bet against Arbuckle? With a landslide of votes, the popular Arbuckle became mayor of the up-and-coming oil town.

Only one formality remained. The man of money, whose daughter's hand Arbuckle sought, happily saw the light and conceded that his daughter had been well won. There was one problem to be dealt with. The West Side, though bountifully endowed with saloons, was not so well represented with churches, particularly ones large enough to hold the multitude of well-wishers who would desire to attend the nuptial ceremonies of a man of Arbuckle's popularity. A

West Side oil fields were the most productive in California in 1913 when *Opportunity* was filmed. Lierly & Son had a fleet of seven tank trucks to help carry oil to railroad tank cars for shipment to market. (Photo from Clarence Williams Collection)

problem-solver to the end, Arbuckle handled this final situation with typical aplomb. The wedding was held, with plenty of rice and old shoes, on the steps of the office at Standard Oil Company's main camp.

It all added up to the dramatic kind of story one is not likely to forget. With one exception, it had everything: true love between a penniless but worthy young man and a beautiful and wealthy young lady, a demonstration of outstanding personal courage in the face of the oil fields' greatest terror, a success story that was phenomenal even by oil field standards, a rousing political campaign, a joyous wedding. All that was missing in the heartwarming story was veracity. Young Arbuckle, in real life, was the well-known actor, Fatty Arbuckle. The stirring events he portrayed were only make-believe. They were part of the script of a moving picture that was, according to its makers, the first feature film ever to be made in the oil fields. The movie was titled, fittingly, *Opportunity*.

There was probably a certain inevitability that some enterprising director in the burgeoning film industry would seize on Taft and the West Side oil fields as a locale for the unfolding of a dramatic story of life in the oil fields. A director looking for a vivid background could hardly have come to a better place.

The West Side fields were the center of the oil universe. They were the most productive fields in California in 1913 when *Opportunity* was filmed. The state, in turn, was the number one oil producing state in the United States. On the West Side, the principal fields were the Midway and the Sunset, lying to the northwest and southeast of Taft. Midway-Sunset produced more than one out of every three barrels of oil produced in California, as well as an amount greater than the combined production of the states of Texas and Louisiana. Midway's wells put out 72,760 barrels a day, Sunset's 12,800 barrels a day, or a total of 85,560 b/d. The closest competitor was the Coalinga district, whose wells put out 53,200 b/d. The more than 85,000 b/d of oil from the West Side was a significant part of California's output of 238,400 b/d. Other leading oil producing states were Oklahoma, 142,500 b/d; Illinois, 76,700 b/d; West Virginia, 30,100 b/d; Texas, 28,800 b/d; and Louisiana, 27,400 b/d.

Midway-Sunset in 1913 produced more than one out of every three barrels of oil produced in California, thanks to wells like this K.T. & O. gusher near Maricopa. (Photo from Clarence Williams Collection)

Flow of oil from newly completed well on Section 30 helped make West Side fields the center of the oil universe. (Photo from Clarence Williams Collection)

It was not just that Midway-Sunset wells produced more oil. They did it in a spectacular manner. The Lakeview gusher flowed more oil in a shorter time than any well had done before, spouting a stream that was dubbed the trout stream and handled with an extensive system of sumps designed to keep oil from reaching Buena Vista Lake. A Kern Trading & Oil gusher four miles northwest of Taft caught fire and lighted the community brightly for more than two weeks, sending a flaming torch into the sky that could easily be seen from a distance of more than twenty miles. An Eagle Creek Oil Company well near Fellows hit a gas pocket and covered the ground for several hundred feet around with fossil sea shells, drawing a crowd of sightseers who gathered shells and marveled that it was possible to do so more than one hundred miles from the ocean.

The trading center for the Midway-Sunset district was the city of Taft, which, in December 1912, a Kern County Grand Jury investigating committee described as "perhaps the livest town in the state." The committee expressed itself as "satisfied" with the situation in Taft.

The Grand Jury's clean bill of health for Taft was described by the *Daily Midway Driller* as "a matter of self congratulation to resi-

Shipment from Anheuser-Busch, St. Louis, Missouri, helped slake thirst in Taft. (Photo from Clarence Williams Collection)

The Old Distillery Saloon on Main Street in Taft. Left to right, Bill Greeson, Chick Brown (sitting on barrel), Ing Simkin (sitting on car, which is a Hudson), and Al Scott. (Photo from Phil Witte)

dents of the West Side and a compliment to City Trustees and local peace officers."

The editorial writer went on to say, "Taftians have no use for 'blue laws' but pride themselves on their good fellowship, boosting proclivities, hospitality and a desire that everyone should enjoy themselves free from undue restraint as long as they observe the proprieties and laws of the state and county. May it ever be thus."

As for good fellowship and the matter of enjoying themselves, Taftians had no lack of convivial places where a man might pass the time. There was the Old Distillery Saloon in the 400 block of Main Street, A. I. Scott, proprietor, H. B. Scott, manager, where the fare included—"drawn from the wood"— Old Jordan, Dant's, Belmont, Old Corn, Old Crow, Hermitage, Carstairs No. 1 and Carstairs No. 6. Also available were Old Distillery bar whiskey, eastern and local bottled beers, Lemp's Famous St. Louis beer on draught and "high class" wines and cigars. For those indisposed to go out, the saloon advertised that telephone orders would be delivered.

Another popular saloon was Robinson's Bar on Main Street near Fifth. Proprietor S. H. Robinson in one breath called on potential customers to "Patronize home industry—by calling at Robinson's Bar for Lyon brew—a Bakersfield beer." In the next breath he advertised, "Just received, 50 barrels of fine wines—Sherry, Angelica and Port direct from the Roseville Winery. Absolutely pure. $2.00 the gallon."

R. E. Woods, whose place of business was on Center Street, advised, "You will meet your friends at Woods' Bar for they know the best brands of wines and liquors are on sale there. We will treat you right." E. J. Boust, proprietor, advertised that fine wines and liquors were available at the Eagle, which also was "auto headquarters for all points in the district." The Schlitz had "that good Schlitz on draught," and J. C. Ketchum and W. H. Fortine told customers to "Drop in and see us" at the Pilot. Of the B & R Thirst Emporium, Curly Chambers, manager, advertised simply, "That's all."

Though Taft had its share of saloons, the drinking man did not always receive encouragement, particularly in the neighboring community of Fellows. When entrepreneurs sought permits from the County Board of Supervisors to open four saloons to augment the two already in business in the community, a petition was presented to the supervisors bearing the protests of representatives of Associated, Pyramid, Midway Pacific, Columbus, Hawaiian, Winona,

General Petroleum, Commodore and Reward Oil Companies. At a hearing, W. J. Atwood, of the Santa Fe lease, served as spokesman, stating, "No big industry was ever developed by booze. We don't need any saloons, we don't want them and we are not going to have them if we can help it." The supervisors turned down the requests for permits.

Though the West Side oil fields were remote from population centers, they could hardly be described as provincial. Oil men traveled the world. Drillers Jack Barrett and W. Fullerton were off to Tampico, Mexico, on a job. Driller Jack Bertol wrote from Trinidad, where he was working for Trinidad Oil Development Company, that the first well had made only 25 barrels a day, adding that the Parker Mogul rotary he was using was the same as those in use in the Midway field. Other West Side drillers joined him in Trinidad, among them Tom Dickey, J. R. Petty, Tom Fair, W. C. Jones, V. B. Cowan and J. C. Gallagher. Rigbuilders L. B. Stitzinger and A. Hitt signed one-year contracts for work in Rumania, where, according to the *Driller*, "It has been found a hard task to find rigbuilders in that cold, bleak country that can erect derricks of sufficient strength to withstand the heavy work necessary to rotary drilling." Drillers Charles V. Reynolds and J. Bailey drifted back from Argentina, disgustedly stating they had been promised $250 in gold a month only to find in Buenos Aires they were to be paid 350 pesos a month, equal to $142 in United States coin. They said they were glad to be back and that "the Midway looks good to us."

In Taft, everything was new and on the way up, so new, in fact, that the *Driller*, not to be outdone by big city counterparts, featured its own down-memory-lane column. Big city papers printed their items under headings that read ten, twenty, thirty and forty years ago; the *Driller* headed its column of bygone days in Taft, "Three Years Ago."

To this setting a filmmaker named F. J. Martin of Film Exhibitors Company came early in January 1913, with a cast of 35 persons, including the aforementioned young Arbuckle, to film the 132-scene moving picture, *Opportunity*.

For West Siders, the filming provided a fascinating chance to see Hollywood stars and a Hollywood movie crew in action. For some, the advent of the first filmmakers on the West Side afforded an opportunity to become bit players and extras. The production hands at Standard's Section 14 camp, immediately west of Taft,

It was in a drilling crew like this that Fatty Arbuckle played a role in the movie *Opportunity*. Real life crew on Standard Oil Company's Section 20 in 1910 included, left to right, W. F. Street, an unidentified man, M. K. Queen and Len Little. (Photo from Jack McCall)

were for purposes of the film given a chance to vote in a mock election. Like many voters before and since, they were not as fully briefed as they should have been. While the camera whirred, they cast their ballots, happily thinking they were electing young Arbuckle to be governor of California. It was not until the picture later played Taft that they learned they had merely elected him mayor of Taft. Other scenes were shot in and around the camp, including pictures of Arbuckle posing as a teamster, handling a team of eight horses lined up with a one-jerk line, and other pictures of him working with the big auto trucks used in transporting casing and rig materials.

On the second day of the two-day shooting schedule, the real excitement unfolded on the McNee lease on Sec. 36, 31S-23E, two miles north of Taft in the area known as Thirty-Six Hill. In prepara-

Street banners featured real-life political campaign in Taft, where Hollywood's Fatty Arbuckle won the mayor's race in the movie *Opportunity*. (Photo from Clarence Williams Collection)

tion, a sump had been filled with an estimated 3,000 barrels of oil. The oil was the heavier variety, which at the time brought 35 cents a barrel, rather than the 31 degrees or higher variety that commanded a price of 80 cents a barrel. Cameras were set up, and the sump was set ablaze. The towering column of black smoke proved only a warm-up for an even more exciting scene.

A wooden derrick had been constructed on the McNee property. A two-inch line had been laid to the center of the derrick floor. Connected to the line was a huge tank filled with crude oil. A pump supplied the pressure to move the oil through the line. Producers of the moving picture carefully moved spectators away from the wooden derrick for, like oil operators, they wanted no casualties. Cameras were positioned. The pump was set in motion, and a stream of oil bubbled up out of the pipe, then shot higher, finally clearing the top of the derrick, rising high into the sky. A daring movie technician slipped up close enough to set the stream of oil ablaze. Flaming oil shot up inside the wooden derrick. A black cloud rose fearsomely above.

To make the scene as realistic as possible, a man at the pump regulated the pressure to permit the oil to flow by heads, allowing the stream to rise and fall in the derrick, thus causing the flames to shoot high in the air and break like a skyrocket, then subside until there was little more to see than smoke.

Against the background of burning sump and flaming gusher, Arbuckle, at a reasonably safe distance from the holocaust, fit a pipe wrench to a length of pipe and began turning. The camera angle made it appear he was performing the chore in the middle of the conflagration.

When film shooting ended, the movie crew left behind a demolished wooden derrick, an oil-spattered hillside, a pall of smoke and a host of excited movie fans, eagerly anticipating the screening of the forthcoming moving picture.

Late in March, some two months after the picture had been filmed, F. J. Martin returned to Taft to make arrangements for the showing of *Opportunity*. The news of his arrival shared space with news of disastrous floods in the Ohio and Mississippi valleys. Flooding was particularly bad in the state of Ohio.

The two events—the flood in Ohio and the arrival of Martin in Taft—suggested a natural arrangement to the open-hearted citizens of Taft. The plight of people in Ohio cities many times the size of Taft stirred an immediate and sympathetic response in the West Side oil community. Why not make the showing of the movie *Opportunity* a benefit performance with the proceeds to go to the citizens of Ohio?

One of Taft's city commissioners, J. W. Ragesdale, who was co-proprietor of the Hotel Alvord at Center and Fourth Streets, advertised as "The Oil Men's Headquarters," lost no time wiring the Governor of Ohio:

"Our people sympathize with you in your sorrow. Will do our part in assisting by sending funds not later than April 5."

In response, Governor Cox of Ohio wired Ragesdale:

"Wire received. Thank you. Half million people homeless tonight. Wire funds to Col. M. I. Wilson, Ohio Flood Relief Committee, Columbus, Ohio. Governor Cox."

With the movie assured as the benefit attraction, the next problem was deciding on a theater. There were two available: the C & C Theater and its rival, the Midway, which had opened its doors some three months before with a free feature length film. The C & C, which had been constructed at Fifth and Center Streets soon after Taft was incorporated, and which furthermore had been the first and only theater constructed in the town with a sloping floor,

Taft's $50,000 Conley School, circa 1912 in this photograph, was a backdrop for part of the action in the moving picture, *Opportunity*. The film's star, Fatty Arbuckle, visited the school to hand out presents to the children as part of the actor's campaign, in the movie, to become mayor of Taft. (Photo from Kern County Museum)

was somewhat disdainful of its rival, advertising "Three new reels of licensed first run pictures every night that do not jump or wobble. See ours then see others."

The Midway, for its part, simply advertised that it had "exclusive first run pictures, changed daily," pointedly including the information that its prices were five cents for children, ten cents for adults. At the C & C, the price of admission was fifteen cents for adults, ten cents for children, and a quarter for reserved seats on the center aisle.

The C & C won. A. T. Connard, the manager, cited the theater's good record in raising money for worthy causes, the most recent endeavor having been the bringing to town of the Pomona Glee Club in a fund-raiser for the Presbyterian church that cleared nearly $100.

The *Driller* proudly announced the showing of the benefit movie on Monday and Tuesday nights, March 31 and April 1, with two showings each night. The newspaper said, "A number of Taft people are to see themselves as others see them when the moving picture entitled 'Opportunity' is to be shown on the West Side. The movie is one taken at the Standard camp about two months ago. The subject is an interesting one, showing the rise of a young man from the

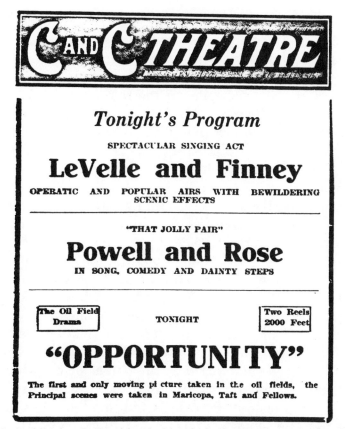

From the *Daily Midway Driller*, March 31, 1913. (Beale Memorial Library, Bakersfield)

position of flunky to that of mayor of Taft and superintendent of the Standard camps."

To enhance the evening's entertainment, the C & C announced it would throw in added attractions. Along with the two-reel, 2,000-foot movie, described as "the first and only moving picture taken in the oil fields, the principal scenes were taken in Maricopa, Taft and Fellows," the theater advertised that the benefit performances would include the "spectacular singing act, Levelle and Finney," whose repertoire featured "operatic airs with bewildering scenic effects," and another duo, "that jolly pair, Powell & Rose, in song, comedy and dainty steps."

On the night of the premiere performance, Taftians filled the C & C Theater to capacity to see themselves as others saw them. The first recognizable picture was of the long rows of derricks in a straight line in the Sunset field on the properties of the Ethel D,

Kern Trading & Oil, Wellman and Monte Cristo Companies. Then came the city of Maricopa with the Gate City Hotel, Oil Well Supply Company and Lakeview Hotel all prominent.

Action moved to the floor of a drilling rig and a cheer went up from the crowd. The man playing the role of Frank Whitney, the Standard superintendent, was Cy Bell, who in real life actually was the Standard superintendent. Bell walked to the edge of the derrick floor on Section 36 and took Arbuckle's application for a position. Cal Wilson, who was working on the rig at the time, also was easily recognizable.

Understandably, the scenes of burning sump and flaming gusher that followed Arbuckle's advent as driller provided the audience with high excitement. As a gusher, it was later agreed, the wild well was realistic to an extreme. Some thought it most clearly resembled the famous Lakeview gusher; others opted for the Pacific Crude wells; a few for the K.T. & O. gusher of a few weeks before.

When the main camp was shown, with Arbuckle as a teamster, the smiling face of the head rooter of the ball team turned out to be Eddie Cronk, seated on a drum of distillate with his collie beside him. Frank Shell was identified as the driver of a six-horse team.

Chauffeur McLaughlin was driving the auto when Arbuckle visited Taft's Conley School and distributed small presents to the children. Many of the children recognized themselves as they filed from the school. The teachers and Principal Hamilton were also prominent. At the ball game, Elmer Woods, Brick Devereaux, Sam Ferguson, Stanton Greene and Merrill McFadden were to be seen going about their usual duties on the diamond.

C. W. Fox of Smith Brothers, the West Side's pioneer clothiers, was one of the prominent figures in the scene depicting the campaign for the office of mayor of Taft. His principal duty, the *Driller* reported, seemed to be to chase "Happy," Tommy Pettit's dog, from in front of the camera.

The grand finale was the wedding scene. On the steps of the Standard Oil Company office, young Arbuckle, handsomely dressed in a formal black suit, white shirt and black bow tie, claimed in marriage the fair Pauline, whose formal skirt ended where the tops of her shoes began. The wedding party, formally and elegantly attired with men in dark suits and ladies in floor-length dresses, included a number of well-known and easily recognizable Taftians, among them Mrs. Walter Fitzmartin, Mrs. Alma Aldrich, Mr. and

C & C Theatre at Fifth and Center Streets served as setting for the benefit showing of the moving picture, *Opportunity*, filmed in the West Side oil fields. (Photo from Clarence Williams Collection)

Mrs. Stanton Greene, Mrs. Tibbet, Mrs. Waitie Parker and Mr. and Mrs. Tommy Pettit.

The moving picture played to enthusiastic crowds each of the two nights it was shown, selling out at each performance and raising more than $1,000 which was quickly dispatched to flood-ravaged Ohio.

If Taft's C & C Theater had been the Broadway stage, there presumably would have been a vigil while producers of the film repaired to the nearest bar to await the verdict of the critics. The producers would have had a good choice of saloons from which to choose, and ample time in which to await the decision. But, of course, Taft was not Broadway, the capacity crowds had been amply rewarded in seeing themselves and their friends on the screen, and *Opportunity* and its owners were off and on the road for showings elsewhere before the verdict came in. The film proved, if nothing more, that Hollywood, in making its first feature movie in an oil setting, had made an entertaining film. However, the story left veterans of the oil fields more than a little puzzled.

The *Driller* in its April 3 edition summed up what amateur critics had to say:

"The oil field pictures, which have been shown at the C & C the past two days, have caused some comment among oil men because

of their being backward in many particulars. It is evident that the
author of the story was far from being conversant with oil field
details, although the pictures would 'go' anywhere they are shown.
Among the men of the fields there are many critics and their notice
was drawn to the picture in several instances because of discrep-
ancies.

"One most noticeable in the rise of young Arbuckle was the fact
that he was given a position as a driller, which means about eight
dollars a day and board. Then he brought in a gusher. He is shown
at this point with a 'trimo' (a pipe wrench) supposedly screwing on
a joint of pipe to control the flow but instead was backing it (the
pipe) off. Had he really made more than one or two twists, it is
evident that he would have been well covered with oil.

"From this position he 'rose' to that of pipe foreman, but the
salary was not taken into consideration. There, in real life, he would
get about three dollars and a half a day and board. Later he is
pictured as a teamster. The rise, in the eyes of the oil boys, was a fall
instead.

"But even at that," the newspaper concluded in a conciliatory
note, "there were none who could say that it was not a big boost to
the West Side oil fields."

West Siders were enthusiastic moviegoers. Capacity crowd at C & C Theatre for
an earlier performance, above, was duplicated at the benefit showing of *Opportunity*, starring Fatty Arbuckle. (Photo from Clarence Williams Collection)

Life in an Oil Camp

As the oil boom spread, eager men built wooden derricks with more alacrity than they showed in building houses. In the wake of an oil discovery in the Reward district seventeen miles northwest of Taft, an oilworker named Bushnell who wanted to bring his family out from Bakersfield solved the housing shortage in his own fashion. He appropriated a derrick that stood over a proposed well which for one reason or another had not been drilled, nailed boards around the floor working area, put a roof over it and lived there with his wife and children for more than a year.

It was common practice following the discovery of oil in a new district such as Reward for owners of leases to provide bunkhouses for single men and, in time, houses for married men who wished to move in their families. The collections of company houses and bunkhouses were known as camps. Each camp normally included a cookhouse and a recreation hall, often known as the clubhouse, as well as an office and supply yard.

One early resident of a Reward bunkhouse was C. M. Small, a rigbuilder for Associated Oil Company, who traveled to the West Side on a hot June day on the "McKittrick Flyer," the long train out of Bakersfield that came crawling once a day up the grade from Lokern to McKittrick. The train was made up of a string of empty tank cars, freight cars loaded with lumber, pipe and oil well supplies, and a combination express and passenger coach. The train had left Bakersfield at 6:50 A.M. and was due at McKittrick at 11:00 A.M., though it was closer to 2:00 P.M. when it arrived. At McKittrick, Small and others were met by the company hurry-up man, who loaded the newcomers' bedrolls and other personal effects on the hurry-up wagon, a low-wheeled buckboard drawn by two long-eared mules, and headed north through hub-deep dust for the Associated lease three miles away. At the lease, Small found the bunkhouse to be a "rough board shack, full of cracks, furnished with homemade bed frames, a homemade table, chairs and a stove—a piece of twelve-inch stove pipe casing heated with gas."

Business center of Reward, 1915, included a post office, cigar and notion store, barber shop, ice cream parlor and general merchandise store. (Photo by O. C. L. Witte)

Along with lodging, the company provided meals. After exposure to the cookhouse, Small said, "We always turned back the top pie crust and looked for foreign objects that might scratch on the way down."

At Kern Trading & Oil Company's bunkhouse in the Reward district, one of the fixtures was a tame crow, which wandered in and out of men's rooms at will. While the men generally liked the crow, they were not so generous in their outlook toward each other, mainly because of a rash of petty thievery. In the absence of any solution to the thefts, everyone was suspect, at least by his fellows, and an aura of suspicion hung over the bunkhouse, reaching its peak when one man lost a gold watch and chain. Fortunately, before any blood was shed, one of the men saw the crow fly out of the bunkhouse dangling a sock from his beak. The crow flew into a large clump of sagebrush, reappearing moments later without the sock. The man investigated and found a cache of loot, including not only the missing watch and chain but small coins, tie pins, buttons, clothespins, a teaspoon and other items.

On the C. J. Berry Oil Company property, there was a unique bunkhouse. Smaller than most, it consisted of a wooden frame covered with wire screen and gunny sacks. On top there was a double roof with a foot of air space, topped off on the outside by a water-filled trough. Water dripped from the trough onto the gunny sack siding, keeping the siding damp at all times. It was the same principle, only on a larger scale, as that utilized in the desert coolers that residents of Reward used for ice boxes in the days before ice was available. The bunkhouse stood on a knoll apart from other buildings, where it might catch any breeze. It provided summer sleeping quarters for night shift men, giving them a cool place to sleep during hot days when they might otherwise have found it difficult to rest.

Since there was no lack of space, the houses that various companies built for married employees in the Reward community sprawled over hills and valleys with only the loosest kind of grouping. Houses adhered to lease camp architecture, consisting for the most part of frame houses with up and down board siding over which battens, or narrow wooden strips, were nailed to seal the joints against wind and dust. Screened porches were part of most houses, furnishing a place to sit and talk and often in the heat of summer a place to sleep. Rent commonly was $5.00 a month for a five-room house, including piped-in water and gas. At Reward, the better

Tenthouses helped solve the housing shortage in early oil development at Reward. (Photo from Kern County Museum)

Oilworkers by cookhouse, State Consolidated Oil Company, Reward, 1915. (Photo by O. C. L. Witte)

frame houses by choice or accident claimed the crestline of the hills, standing in treeless majesty along what came to be known as Silk Stocking Row. On the slopes were more modest batten houses, "rag houses" fabricated of discarded canvas and lumber, and tents that earned for the real estate on which they stood the sobriquet of Rag Gulch.

Initially, food and supplies were hauled into Reward by wagon or buggy from McKittrick. As more families moved into company houses, a general merchandise store was established to serve the sprawling community. The store contained quarters for a post office, which became a daily gathering place. Another store followed, offering sundries, hard liquor, magazines and jewelry, among other things. The latter store, built by M. P. Scott, had a cardroom in the rear. No law established closing hours, and the cardroom stayed open as long as a game was in progress.

A popular addition to the tiny business community came with construction of a lean-to on the west side of the original store, owned by Gordon D. Christian. The lean-to served as an ice cream parlor. Miss Irion Bass, whose father worked for Kern Trading & Oil Company, presided over the ice cream parlor, preparing sodas with fizzwater kept in a barrel in the cellar. When Christian decided he needed the space for expansion of his store, he sold the ice cream "franchise" to Ralph Agey, who built an addition on the other side of the Christian store to serve as an ice cream parlor and barber shop. Agey served as barber. His wife ran the ice cream parlor.

Life in the oil camps that formed the community of Reward was to a degree more advanced than in cities. Housewives enjoyed the convenience of piped-in gas at least five years before it was available in Los Angeles, and at no cost. Gas, a by-product of oil production, was in plentiful supply, and it was furnished free of charge to residents of the community. While women in Los Angeles used coal and wood in stoves, the housewives of Reward baked cakes in gas-fed ovens, getting results that were widely appreciated, particularly when the cake found its way out to a cable tool drilling rig for an evening of family socializing mixed with work. Gas not only furnished a source of fuel for cooking and for heating houses but also furnished light. Gas lights, consisting of mantle, globe holder and globe attached to a gas line, furnished illumination that could be controlled from dim to bright by adjusting the amount of gas fed to the flame. Gas also furnished fuel for a common household appliance, the iron. A flexible hose joined the gas source with the iron,

Supply wagons entering Reward, circa 1913. (Photo from Ann Bass McDonald)

delivering gas to the iron to be burned to create heat that could be regulated to provide a warm, medium or hot iron, as desired. Because the flow of gas from wells was erratic, flaring from time to time, it was necessary to install a regulator on the gas line outside the house. Generally called a gasometer, the regulator prevented any sudden flare-up that might endanger the house or its occupants.

Another convenience enjoyed by residents of Reward was a ready supply of hot water for use in the kitchen and in bathing. Steam furnished the power to run engines at oil wells, and the countryside was laced with lines carrying steam from central boiler houses. For an oilworker, it was no job to tie into a steam line, piping live steam to his house. Outside, the steam tap would drip into a tank set on a platform perhaps six feet high. The condensed water was very hot, and it was piped into the kitchen sink and to the bathtub. Along with being hot, the water was as soft as anyone might desire.

Steam provided another bonus. At each boiler house a pipeline led away to a reasonable distance from the boilers, dead-ending to provide a means of blowing off excess pressure. At the end of the line, there would be a sturdy box with a lid that could be fastened tightly on top. The box and lid were perforated with holes. The blow-off box provided a handy oil field washing machine. The oiliest of work clothes, when soaked in distillate and introduced to a session with live steam in the blow-off box, came out clean.

In one company's residences, the boiler house provided a solution to the scarcity of potable water in the Reward district. Kern Trading & Oil Company furnished condensed water from coolers at its boiler house to serve as drinking water in its lease houses. Twice a week a tankwagon stopped by each house to fill the covered containers that held drinking and cooking water. For general use, the oil company piped in sulphur water, secured from wells west of Reward.

Other companies solved water problems in various ways. C. J. Berry Oil Company had a water well near a spring on company property. Associated Oil Company and others brought in water from wells three miles away in Little Santa Maria Valley. The same wells, known as the Santa Fe wells, provided water for McKittrick.

Though the general merchandise store at Reward stocked a surprisingly large inventory, it did not have space to offer a wide selection of brands or sizes, nor could it offer many perishables because of a lack of refrigeration. Stores in McKittrick, three miles away, were not a great improvement. Serious shopping often re-

Feeding the flock at a lease house, State Consolidated Oil Company, Reward, 1915. (Photo by O. C. L. Witte)

Above, porch swing at a State Consolidated Oil Company lease house, Reward, 1915. (Photo by O. C. L. Witte)

Below, children at play beside a Reward lease house, 1915. (Photo by O. C. L. Witte)

Well-dressed young ladies of Reward, 1915. (Photo by O. C. L. Witte)

Business center of Reward, circa 1912, included post office run by M. P. Scott in his cigar and notion store, barber shop run by Ralph A. Agey, ice cream parlor run by Mrs. Agey, and general merchandise store run by G. D. Christian. (Photo from Kern County Museum)

quired a trip to Bakersfield, forty miles away. It was not unusual for a family from Reward to travel by train to Bakersfield one day, shop, and return the following day.

Another possibility was mail-order purchase from such stores as Black's Package Company of Fresno. The Bush Bass family, among others, availed themselves of this approach, receiving the shipment by rail. A typical order might include a couple of kegs of pickles, sweet and sour, a large wooden box of crackers, some cookies, a case of Vienna sausages, deviled ham, sardines, peanut butter, flour, corn meal, a fifty-pound can of lard, cases of canned vegetables, a ham and a slab of bacon. The ham and bacon would be hung on the back porch, which was protected by morning glory and Madeira vines.

For those who stayed home, there were peddlers to seek them out. Among the peddlers was a Chinese vegetable man, who came through the fields driving a two-horse, double-deck wagon with canvas flaps that covered the sides and back, shielding his wares from sun and dust. His vegetables were shipped to McKittrick, where he picked them up to carry out into the fields. He was a popular weekly visitor, especially at Chinese New Year, when he gave customers gifts of nuts and candies.

Another peddler was a dry goods man known as the Assyrian. Traveling by horse and light spring wagon, he carried his merchandise in huge leather cases, which he rolled into the house for inspection. He handled yardages, threads, trimmings, curtains and table linen, among other items. He also carried sample dishes. If the customer ordered a set, it would be shipped later. The Assyrian's sample dishes were the size of doll dishes, and an order for a standard-size set might bring a gift of the samples to the daughter of the house.

The Watkins man, who passed through at least twice a year, carried such merchandise as spices, extracts, baking powder and liniment. If a customer made a purchase, the Watkins man took it out of stock so there was no wait for delivery. Company cookhouses and bunkhouses were normally open to the peddlers, for companies recognized the inconvenience to employees of living in out-of-the-way communities like Reward and wanted to ease the discomfort by encouraging peddlers to bring necessary wares.

In the absence of a doctor, residents of Reward often weathered ailments with home remedies. There were several practitioners whose knowledge went back into Indian lore. One was Mrs. Mitchell. One of her remedies involved the deadly nightshade. She would gather the leaves of the poisonous plant, dry and crush them, and burn the residue to help clear breathing passages afflicted by severe colds. Another remedy involved a plant known as "dog fennel" which grew in the area. She used the dried leaves to make a medicinal tea to combat diarrhea. For boils, she recommended yellow soap poultices. For bee or wasp stings, wet mud. Once one of the children at Reward suffered a badly cut foot that bled profusely. Mrs. Mitchell seized some spider webs, washed them in the palm of her hand with alcohol and applied the webs to the wound, quickly stopping the bleeding. The cut healed rapidly, without infection.

For children, there was no lack of exciting things to do at Reward, no scarcity of pipelines to be tightrope-walked across canyons, no end of tarantulas to be caught and kept as household pets, no shortage of fun to be enjoyed riding Edna Yeager's donkey, or Phillip Small's. If one had a dime, Edna or Phillip would rent their donkeys for the day. With Phillip's donkey, one might have to get off and lead it most of the way. With Edna's jenny, one might have to pick

Left, well gang at work. Right, boilers furnished steam to power drilling rigs, pump oil and heat crude for shipment through lines. Condensation from steam lines also furnished hot water in lease houses. Reward, 1915. (Photos by O. C. L. Witte)

oneself up from the ground a time or two.

In the spring, there were poppies, Indian pinks, bluebells, daisies, buttercups, primroses, fiddlenecks and cream cups among wild-flowers to be enjoyed. Hidden among wild oats that grew by the spring near the Reward store were Indian arrowheads and beads waiting to be found by those with sharp eyes, for oilworkers were not the first to live where Reward's houses sprang up. Indians had left a fascinating legacy in colorful hieroglyphs painted with enduring dyes on Carneros Rocks in the nearby Temblor foothills.

There were storytellers to be listened to, men like Henry Nichols, a bachelor who had been born in Grass Valley, California and could tell exciting stories of the Mother Lode. He knew stories of stage-coach holdups and mining tragedies, and he could recall the names of bandits and give the details of their escapades. Once one of his listeners, wondering at his command of details, asked Henry if he had been one of the bandits. Henry just laughed.

Another favorite was Mr. Briggs, who was the blacksmith on the Kern Trading & Oil Company lease. He had been a prospector and still liked to hunt for gold. He always seemed to have a supply of gold on hand and generously made rings and stick pins for almost anyone who asked. The youngsters liked to watch him melt the gold and work it into a ring or pin. He had a scar on his cheek which he said was caused by an Indian's arrow.

Formal education was not neglected. About a mile from Reward at the rail terminus at Olig, which took its name from the Greek prefix meaning "scant, deficient," there was a one-room school-house which doubled as a non-denominational Sunday School on Sundays. When the building was being used for classes, a single teacher taught all eight grades, with perhaps thirty to thirty-five pupils. Mr. Winters, one of the first teachers, was a short, heavy man who sometimes had his hands full disciplining older students, many of whom were as large as he, if not larger. One of the bigger boys' favorite pursuits was hiding the hand bell that Mr. Winters used to assemble students after noon or recesses. They would hide the bell in one of the chinaberry trees by the schoolhouse, or in the attic, or on top of the porch, or on the very top of the schoolhouse. Once they hid it on top of a fence post, which was such an obvious place Mr. Winters had more than the usual difficulty locating the bell. His students stood on what they considered their rights and refused to go back to class until he rang the bell. He secured a fifty-pound lard can top and pounded on it as a substitute, but no one paid any attention, waiting until he finally found the bell.

When the long day's work was over...oilworkers at Reward worked a 12-hour day, seven days a week. (Photo by O. C. L. Witte)

Once a year cowboys held a big roundup, driving cattle to the shipping yards at Olig. When cowboys drove the herd past the Olig school, the teacher would let pupils out to sit on the fence and watch the herd go by. The cowboys, wearing jeans and either sheepskin or plain leather chaps to protect them from thorns and cold, would ride up to the fence to visit with the school children, happy to have someone to talk to.

Aside from going to school, there was another reason for Reward's youngsters to trek to Olig. The train that came out from Bakersfield would, after discharging passengers at McKittrick, pull up to Olig, the terminus, and turn around. Friendly trainmen had no objection to the youngsters hitching a ride to McKittrick, free of charge. Though the youngsters might have trouble getting a lift back to Reward and might even have to walk, they would almost certainly be treated to a soda pop by a friendly oilworker or two in McKittrick.

A few miles west of Reward in the Temblors was Maddox Ranch, where a large two-story frame building with a porch all around served in the downstairs portion as kitchen and eating quarters, upstairs as a bunkhouse for cowboys working on the ranch. Youngsters from Reward liked to hike up to the ranch, where they were assured a warm welcome, a free meal and all the fresh milk they could drink.

A mile or two above the oil fields in the Temblors there were sulphur springs from which a good stream of water flowed year round. Water was piped down the hill to a meadow, where it went into long troughs from which cattle drank. A short distance from the watering troughs there was a shallow lake, which filled during spring and held water through most of the summer. The lake was a favorite place to play for youngsters who poled rafts and skipped rocks over its surface. The sulphur springs were also popular with adults. The Southern Pacific lease carpenters built a two-compartment bathhouse, and persons suffering from arthritis and rheumatism "took the baths" in the warm sulphur water.

Perhaps the highlight of each year for children and adults alike was the annual community Christmas party. At either the Reward Oil Company clubhouse or the Associated clubhouse, there would be a community Christmas tree and entertainment in which school children as well as adults took part. The tree, generally a huge one cut by a crew sent back into the Temblors, would be loaded with

The only jobs open to Orientals in the oil fields were as cooks or gardeners. Yip Mai was the Chinese cook at State Consolidated Oil Company's lease at Reward in 1915. (Photo by O. C. L. Witte)

candy and gifts and each child would have one or more gifts, purchased from club funds. In addition, there would be a generous bag of candy, nuts and fruit to make the holiday season a memorable one.

Since most buildings in the Reward district were of wood, fire was an ever-present danger. Preparations to defend against it consisted of placing taps on water lines that ran through the fields so that a hose might be connected in a hurry, also in forming volunteer companies to man two-wheel carts equipped with hoses and chemical fire extinguishers. The signal for fire was a continuing series of short blasts from the boiler whistle nearest the fire. Each boiler whistle had its own pitch, which for the initiated was readily identifiable. At the sound of the whistle, men in the field were to drop what they were doing and rush to the scene of the fire to fight the conflagration.

On the Associated lease, men in supply organized a volunteer fire department captained by D. F. Devlin. They purchased a forty-gallon chemical tank mounted on two wheels, and laid out a contingency plan that called for C. M. Small, former rigbuilder who had been promoted to foreman, to rush to the warehouse at the first alarm to pick up the fire-fighting crew and equipment with his buckboard, which was drawn by two spirited horses. True to plans, the next time an alarm sounded, Small pulled up posthaste in front of the supply warehouse, and the fire crew rushed to tie the chemical cart on behind. The fire fighters climbed into the back of the buckboard, Devlin scrambled in beside Small—and the team ran wild. Men dropped off the back, leaving the chemical cart to careen madly at the end of a ten-foot rope. The tank charged first one way, then the other, until it finally teetered on one wheel and turned over. When the crash came, the buckboard almost turned over too, and Small was thrown out. Devlin jumped. By the time the crew gathered up its equipment, the fire was out.

On another occasion, a young foreman named Henry Cable had stopped by one of the residences, dismounting from his light buggy without bothering to tether the horse. A fire whistle blew on the McKittrick Oil Company lease. Before Cable could get to his buggy, the horse had bolted. Cable followed the tracks down the road, around sharp curves, past two right angle turns, across a bank between two sumps and through another turn which brought the fire into sight. The tracks led straight across the field, over pipelines and sagebrush, to the column of smoke that rose from a general merchandise store near Olig schoolhouse. There Cable found his horse and buggy. The horse was none the worse for wear. The buggy was badly damaged, its seat missing. The horse, Cable learned, was a retired fire horse.

The capital that underwrote development of oil production at Reward largely came from financial centers far removed from the oil field. To men who put money into bringing in wells, the community of Reward sometimes loomed as a place to serve more functions than one, a place not only to invest money but to send sons who perhaps did not seem to be measuring up in the college environment, in short, a place for a youth to find himself. This perhaps was the way the Bowles family of San Francisco viewed the community. Bowles, a financier with an interest in Reward Oil Company, had a son, Phillip, whose performance at college apparently left something to be desired, at least in the eyes of his parents. In an effort to straighten out Phil, the senior Bowles exiled him from the bright lights of San Francisco to the simpler living of Reward. The change shaped up as a major one, from the delights of a sophisticated city to the isolation of an oil camp where summer sun beat down mercilessly on frame houses, tanks and wooden derricks, and cold winter fog obscured them.

To some, the change might have been a sobering one. But Phil Bowles was a large, robust youth, full of pep, fun and vigor. He loved people and animals, he liked to gamble, and he was full of fresh and unusual ideas. In short, he was just the kind of man they needed in the oil fields. Far from being out of his element, he fit right in at Reward and quickly became one of the most popular men around. When the elder Bowles decided his son had served his penance and should be permitted to return to college, young Phil Bowles chose to stay in the oil fields. He accented the decision by marrying Jessie Jackson, whose father was superintendent of San Francisco Oil Company. The marriage was not exactly what Bowles' parents had in mind, having picked a more socially prominent girl in San Francisco, but it proved a happy one.

When Bowles looked at the oil fields, he naturally seemed to come up with ideas. One was a cracking plant to make gasoline from crude oil. He put together a model on the Reward Oil Company property. The model, behind locked gate, attracted much interest, including a visit by a group of Japanese industrialists who drove out to Reward in a limousine and, as Bowles' guests, carefully inspected the plant, much to the excitement and curiosity of onlookers sitting on the steps of the Reward store.

Bowles put in a swimming pool by the bunkhouse. He installed an ice-making machine at the cookhouse and permitted the sale of ice

Picnics in the Temblors were a social highlight of life in the oil camp of Reward, 1914. (Photo from Kern County Museum)

A picnic in the Temblors, circa 1915. (Photo by O. C. L. Witte)

to anyone who wanted it. He equipped the cookhouse with ice cream freezers. He added a poolroom with two tables to the clubhouse and saw that a moving picture screen and a projection room were provided, and a movie shown at least once a week. He had a stage built at one end of the clubhouse with dressing rooms on both sides and made carpenters available to build sets for the shows that were given there. Among featured performers were oilworkers Carr and Lindoefer and their wives, who were ex-vaudeville performers, and a man known as the Old Fiddler who lived alone in a tent, repaired watches and played his fiddle and sang at community parties. The Old Fiddler brought down the house with such favorites as one that went, "Oh vot's de use, Teddy de Roose, he killa de lion, de tiger, de bear, as fast as dey turn 'em loose."

Not all who took up residence in Reward came from such a socially acceptable background as Phillip Bowles, nor had all been known by the names they went by in Reward. One such man was an oilworker who, for the recounting of his story, will be known as Jack Altman. Altman spoke with a Texas accent, as did many others in the fields, but he said little about the old days. When others traveled home on the passes that were given by Kern Trading & Oil in the days when it was owned by Southern Pacific Company, Altman stayed in California. He was a quiet man who was accepted in Reward as a hard worker who minded his own business and did not borrow trouble. His closest friend was a fellow Texan to whom Altman had advanced the money to come to California and who, once arrived, had been loaned money by Altman to see him through to the first payday. When, after a number of months at Reward, Altman's friend had saved enough money to send for his wife and children, it was at Altman's frame house that the newcomers enjoyed their first hospitality. Through ensuing years, the friendship continued until Altman retired, a respected man, and with his wife moved to Northern California. This was the part of the story that many people knew. There was another part that only Altman and his friend knew.

It went back to days when both were young in a small town in Texas. Altman had been involved in a fight in which a man was killed. The law had sought him, but had been unable to find him. His friend, a clerk at the time in a general merchandise store, had known Altman's hiding place—under the store—but had not told anyone. In time, Altman had traveled west to California, where he had taken a new name and started a new life in the oil fields. However, he did not forget the man who had helped him. When he heard there was a depression in Texas, he prevailed on his benefactor to come west. He sent the money to enable him to make the trip. He helped him get a job, and gave him as much support as he could. Such was their friendship that Altman was willing to risk his new reputation, his new life, to help the man who had helped him.

If the past never caught up with Jack Altman, the same could not be said of a lady to be known here as Josie, who lived with her presumed husband, whom we shall call Henry Holbrook, and her small daughter, whom we shall call Angelina. One morning after Holbrook had gone to work, a lone horseman rode into Reward. He seemed to know where he was going, which was directly to the frame house where Josie lived. The stranger dismounted, but before he could enter the house, Josie ran screaming from the back door, shouting that a man was going to kill her. A neighbor hurried to the house to find the stranger standing quietly on the porch. By the look of the man's horse and his clothing, he had been on the trail a long time. However, he did not seem violent. He assured the neighbor that he meant no harm to anyone and said he just wanted to talk with Josie. He said he wanted them to know why he was there. He had been hunting for Josie for almost three years. She was his wife and the little girl, Angelina, was his daughter. Josie had run away with another man, taking Angelina with her. The stranger said he just wanted to see that the child was all right and well cared for.

While the neighbor waited on the porch with the stranger and Josie, another neighbor found Henry Holbrook in the fields. Holbrook took the rest of the day off and through the afternoon Josie, the stranger who said he was her husband, and Holbrook could be seen on the porch, talking, apparently without bitterness, while Angelina played nearby. At nightfall, the stranger was permitted to stay in the Pacific Oil Company bunkhouse. In the morning, after breakfast at the cookhouse, he mounted up and rode away. The incident was closed. Josie, Henry Holbrook and Angelina continued to live in Reward, mingling with others as before, with no questions asked of the past.

Among others who called Reward home was Slim Abrahams, a cable tool driller. One night on the job Slim decided there was

Oilworkers took as much pride in the cars they drove as cowboys in their horses. O. C. L. Witte, hand on radiator, and C. C. Rea, left, jointly purchased this Model T Ford in 1915 for $415.70. Additional cost including shock absorbers ($12.50), cut out ($2.50), chains ($5.00), and license ($2.20) boosted total expenditure to $437.90. (Photo from Phil Witte)

nothing he would like better than a chicken dinner. The forge made a handy spot to fry chicken. Slim's tool dresser, a man named Vershey, told Slim to be patient; he would rustle up some chickens. He left the rig and returned a short time later with three dressed chickens. Slim asked no questions, feeling that the less he knew the better. The two men enjoyed a nocturnal chicken fry. The tour done, Slim returned home and went out to feed three chickens he had cooped up for fattening. The chickens were gone. He never knew for sure, for Vershey was uncommonly vague, but Slim had a reasonably good idea what had happened to his chickens.

There was Lawrence Merriman, who worked on the Southern Pacific lease and was an artist in his spare time, painting sometimes on canvas, other times on pie tins, which he sold as souvenirs of the oil fields.

And Swede John, a husky oilworker who set out on a dark and foreboding night on his motorcycle to travel the torturous three-mile road that led from Reward to McKittrick. It was a narrow, pot-holed excuse for a road that had been churned to dust by hooves of horses and mules and the hard wheels of oil field trucks. Swede John prepared to travel it by drinking himself into a pleasant state of relaxation. Seating himself on his motorcycle, he rode off into the night. He had not traveled far before he saw a blurred but frightening apparition bearing down on him. The apparition resolved itself into what appeared to be four headlights, gleaming balefully out of the darkness. Knowing that cars came equipped with only two headlights to a vehicle, Swede John surmised he was about to get clobbered by one or the other of a pair of cars, one of which obviously was passing the other. He debated whether to take to the ditch. Envisioning the rough terrain, he decided this would not be worthwhile. Perhaps pride seized him. He had as much right to the road as any combination of cars. He decided on a daring maneuver. To save himself from impending disaster, he steered a straight course, meaning to pass between the two cars. When he awoke, swathed in bandages, Swede John learned he had hit dead center between the headlights of the single car that had been his adversary.

The proud city of Taft boasted home-delivered bread and other bakery treats from Cox & Foster's Bakery as early as 1910. (Photo from Clarence Williams Collection)

The Oil Field's
Greatest Sporting Event

For a place that three years before had been nothing more than a railroad siding, the oil town of Taft in the fall of 1913 had every reason to consider itself lucky. For one thing, the oil fields that gave the town its sole reason for existence were booming. Seldom a week went by without announcement of the completion of another big well, like the 3,700 barrels-per-day producer that Northern Exploration Company brought in on Sec 22, 31S-23E, four miles north of town, or the 3,000 barrels-per-day well that Spreckels Oil Company had gotten on its Maricopa Flats property. On the McNee property alone, two miles north of Taft, Standard Oil Company's gushers were, in the words of one observer, "turning more than $20,000 a day to John D. Rockefeller's pocket."

Along with the oil boom, there was a building boom. The Dreamland skating rink had just opened, the C & C Theater was being remodeled, a new schoolhouse was to be built, the Odd Fellows Hall was nearing completion. As if that were not enough, even the prices of oil stocks were moving well. Union Oil Company, which had shown some weakness, was bouncing back, up $11.50 a share in two months—from $51 to $62.50. For those who speculated in penny stocks, there was something to cheer about too. Jade had doubled from four cents a share to eight cents, largely on the basis of the company's land position near a well that was being drilled by another company.

Yet one thing was lacking. An experience related by a Taft resident illustrated the problem. This particular person, an oilworker proud of his part in helping to build a town from scratch, had come to carry sandbags for the dike thrown up to contain the Lakeview gusher and stayed on to become a pipeliner. He had journeyed back to the city where he had grown up, which happened to be San Francisco, where he told those who would listen all about the new town that was booming in the oil fields, of flaming wells and high pay, of tent houses and friendly people. The response was uniformly disappointing.

Gushers like the Lakeview No. 2 helped make the West Side a major oil producer, but the city of Taft suffered an identity crisis. Few outside the oil fields knew where the city was. (Photos from Clarence Williams Collection)

"Taft?" the listener would say. "Where's that?"

If the outside world had not heard of Taft, there appeared in late September in the town's newspaper, the *Daily Midway Driller*, a news story that promised to remedy the oversight. George Wilson, the manager of a young heavyweight named Jack Lester and a sometimes promoter of boxing contests, announced that after weeks of negotiations he had secured the services of the great Sam Langford, the world-famous Boston Tar Baby, for a 20-round match against Jack Lester. The contest would be held in Taft on Monday, October 27.

The significance of the impending battle was not lost on civic-minded citizens of Taft, including those who were not fight fans. If Tex Rickard could make Goldfield, Nevada known throughout the country by staging the Joe Gans-Battling Nelson lightweight championship fight there, George Wilson could do the same for Taft with the Langford-Lester match. "This means the putting of Taft on the map in the matter of fisticdom of the high-up variety," wrote a newsman in the *Bakersfield Californian*, "and to George Wilson belongs the credit." Noting that Wilson had wired money to Langford's manager for three tickets from Boston to Taft, the newsman continued, "Wiring of tickets meant putting up a big guarantee, but to the city of Taft it is worth more than double the amount in advertising the city will get from the big go."

Taft's Blaisdell Opera House featured such entertainers as the Hoozier Bloomer Girls and Dad Sloan's Fifties, circa 1913. (Photo from Clarence Williams Collection)

A home talent show attracted a capacity crowd to the Blaisdell Opera House, circa 1913. Among those taking part were, front row, Annie Dougherty, second from left; Martha Ogle Dougherty, third from left; Mrs. Kellermeyer, sixth from left; Mrs. Canfield, whose husband was superintendent of Chanslor-Canfield Midway Oil, seventh from left; Mrs. Guiberson, ninth; Mrs. Fred O'Brien, 14th; Mrs. Tupper, 16th; Mrs. Duff, whose husband ran a drug store. Back row, Mr. Kellermeyer, sixth from left. (Photo from Clarence Williams Collection)

No mention was made of the irony of selecting Langford, a Negro, as the man to put Taft on the map. Negroes, according to an unwritten law, were not permitted to stay overnight. Nevertheless, the contest shaped up as an interesting one. Sam Langford was the number one contender for the world's heavyweight championship. The champion was Jack Johnson, who was living in Paris. Langford had met Johnson seven years before, in 1906, losing a 15-round decision that was widely disputed. He had been chasing Johnson ever since for a return match, particularly since 1908, when Johnson dethroned Tommy Burns for the title in a Christmas day fight in Sydney, Australia, and with renewed vigor since 1910, when Johnson scored a 15th round knockout over Jim Jeffries, who had retired several years before as undefeated champion. The latter victory in a Fourth of July contest in Reno, Nevada had clinched Johnson's claim to the world's championship. Langford, described by one boxing writer as one of the greatest fighters that ever lived, had campaigned with great success in eastern rings and in Australia and Canada. The match against Lester would be his first in California in four years.

Lester, who was unrated, was a relative newcomer to Taft, having come to the oil fields five months before, a strong young man looking for a fortune. But being new did not mean anything in Taft. The town itself was only a few years old, and everybody there was originally from someplace else. In Lester's case, the someplace else was Cle Elum, Washington. A shade under six feet tall, Lester was likened by some to a larger Battling Nelson. To his credit, he had gone 20 rounds with Sam McVey, the world's 12th-ranked heavyweight, and though he had lost, he had acquitted himself well, particularly in view of his lack of experience. In his initial appearances in the ring in Taft, he had seemed awkward, though willing. There could be no doubt of his courage, though there was some question of his dedication to proper training. Outside the ring he had acquired something of a reputation for high living. But he had won a local following and in his last outing, he had held the hard-punching Kid Kenneth, whom the *Driller* described as "the cleanest sport and the best fellow that ever came to town," to a 12-round draw. There were many who thought Lester should have had the decision. There was little likelihood of a rematch, however, for as George Wilson put it, "Jack has taken a wonderful liking to Kenneth and does not take well to the idea of meeting him in the ring again."

Lester was only 22 years old. He was improving with each appearance and was obviously on the way up. The match with Langford, the number one contender, would give him his big chance.

If the city of Taft had deliberately set out to choose the most propitious time to make a play for the attention of the sporting fraternity and with it the attention of the nation, it could not have chosen better. Even as the town basked in the reflected glow of the coming contest, situations were deteriorating elsewhere.

In Bakersfield, Taft's rival and many times its size, the citizenry could not look forward to seeing a big-time contest like the Langford-Lester match. They were having enough trouble seeing matches pitting relative unknowns against each other. A scheduled match between Kid Dalton, an Italian from Los Angeles, and Kid Booker, a Negro from Bakersfield, and lightweights at that, had been run out of town. At the instigation of the Law & Order League, the principals had been arrested and hauled before a local judge. The judge had promptly released the two lightweights, though the match itself had not been allowed. Bakersfield had to suffer the ignominy of seeing the match go to, of all places, Maricopa, where Promoter Frank Fields had it scheduled for the Majestic Theater. It was small consolation that special stages would be run to carry Bakersfield fans to Maricopa for the battle. The best Bakersfield could offer in entertainment could hardly be compared with something as outstanding as a chance to see the number one contender in a match

Taft Ice Delivery in 1913 guaranteed "none as good as Crystal Distilled water ice." (Photo from Clarence Williams Collection)

that might produce the next heavyweight champion. It was a touring company that would soon present on the stage of the Bakersfield
Opera House the play, *The Traffic*, described as "a compelling
expose of the actual fact of white slavery as it exists in the cities of
the nation."

Nor were things going much better in Los Angeles, the metropolis
to the south. There Promoter Tom McCarey was having trouble
trying to stage a lightweight contest between Johnny Dundee and
Joe Azevedo. The Los Angeles Church Federation had secured
15,450 signatures on petitions demanding the outlawing of boxing in
California. The group was putting heavy pressure on the district
attorney to stop the Dundee-Azevedo match, which Promoter
McCarey discreetly chose to advertise as a "sparring exhibition,"
offering free seats to any clergyman who wanted to attend. The
harried district attorney, in response to church pressure, promised
to have deputy sheriffs at ringside at Vernon "to censor the contest,"
adding, however, that "unless the bout assumes the aspect of a
prizefight in the opinion of the deputy sheriffs, there will be no legal
interference." It was hardly the kind of advance publicity that would
sell tickets to fight fans who wanted to know which of the lightweights was the better man.

Even San Francisco was having its difficulties. In a stroke of what
could only be described as questionable timing, the city was in the
process of eliminating Barbary Coast as a prelude to staging what
was designated as the Portola Festival, a four-day celebration com-
memorating the 400th anniversary of Balboa's discovery of the
Pacific Ocean. As part of the great celebration, numerous units of
the Pacific Fleet would put in at San Francisco, among them the
cruisers *Charleston, St. Louis, South Dakota* and *Pittsburgh*; the
gunboat *Yorktown*; the supply ship *Buffalo*; and the destroyer *Far-
ragut*. The latter vessel would carry the man chosen for the role of
Balboa through the Golden Gate for a festive docking in San Francisco, to be followed by a ride at the head of a parade up Market
Street. It seemed doubtful that men aboard the Navy vessels had
been consulted in advance with regard to the city's decision to shut
down Barbary Coast, but the city was proceeding anyway. The
newspapers that carried the story of the signing of Sam Langford to
fight Jack Lester in Taft also carried front page stories describing
the last nights of Barbary Coast. The people who were closing it had
picked an effective means. They had simply banned the sale of

Strong men forged a booming oil industry at Taft, and professional boxing was one of the biggest sports they paid their wages to watch. The Sam Langford-Jack Lester heavyweight contest stirred keen interest. (Photo from Clarence Williams Collection)

liquor. It was estimated that 500 "girls" would be thrown out of work by the closing of the Coast's dance halls. The San Francisco Women's Club announced the opening of a headquarters on the edge of Barbary Coast to assist the girls "in finding employment other than that to which they have become accustomed."

If others did not know how to run things, entrepreneurs were not backward in Taft. No sooner had the fight been signed than it was announced that construction was starting on a great new pavilion which would be inaugurated by the Langford-Lester fight. The $10,000 structure, to be built on Fourth Street adjacent to Recreation Park, would have a seating capacity of 6,000, or more than three times the number of residents of Taft. It would include a 24-foot square ring—the regular Marquis of Queensbury size—as well as a specially manufactured hand-woven canvas so thick that should a fallen man's head strike the mat, no harm would result; it would, the *Driller* reported, "be like landing on a heavy Brussels carpet."

To be built by R. & W. Amusement Club, Incorporated, whose directors were George Wilson, S. H. (Pop) Robinson and S. J. Dunlop, all of Taft, the pavilion would serve as a setting for what the backers described as all sorts of amusements—boxing contests, theatrical peformances, dances and other amusements "of a moral and legitimate nature." Gas and electricity would be installed for

Like dignitaries before him, Sam Langford, the world's number one heavyweight contender, arrived in Taft on the Sunset railroad. Crowd above was on hand to greet excursion visitors from San Francisco who came to see the oil fields circa 1910. (Photo from Clarence Williams Collection)

year-round use. In the summer, the octagon-shaped pavilion would have a scenic railroad, a chute-the-chute and a plunge. The plunge would be filled with a mixture of salt and sulphur water to be obtained from wells in the oil fields. The two waters, when mixed, were said to provide delightful bathing. Such bathing, the *Driller* noted, "has been experimented with here and results have been found to be beneficial for the skin as well as a curative for rheumatism and similar complaints."

Even as the coming contest gave the outside world something with which to identify Taft, it gave men who worked in surrounding oil fields something new and exciting to discuss. Most talk centered on Lester's chances. There were a few men who had seen Langford fight, and they were inclined to string along with him. Others argued that Langford, at 29, was over the hill, that Lester was on the way up and that he was the man to bet on. One telling point for those who argued Lester's chances was the news account of Langford's last fight, which had been a 12-round main event in New York's Madison Square Garden against Joe Jeanette. Langford, according to the account, had come in at 199½ pounds. If true, it meant he was 15 pounds or more over his best fighting weight, which normally ranged from 180 to 185 pounds. Perhaps age was beginning to take its toll.

Talk of the big fight crowded out other topics, like the excitement at Edison and in Cuyama Valley. At Edison, near Bakersfield, George Hay and Associates had been drilling a water well on Sec. 6,

30S-30E. At a depth of 400 feet, they had made a startling discovery, encountering oil sand and gas. The excited water well drillers had quickly changed plans and decided to go down another 1,000 feet. First, though, they had had the foresight to take options to lease four thousand acres in the vicinity of the well.

In Cuyama Valley, a favorite hunting ground for Taftians some 15 miles by road out of Taft, someone had been poking around the bed of the Cuyama River, presumably a hunter, and had stumbled on a live oil seep making an estimated half a barrel a day. News of the find had sent a horde of men hurrying over Grocer Grade to Cuyama, where in less than a week's time they located more than sixty mineral claims in anticipation of the day when Cuyama would be the next big oil field. As exciting as the discoveries in Edison and Cuyama might be, neither could hold a candle to the coming fight.

While Sam Langford and his party began the long train trip from Boston, Jack Lester departed for California Hot Springs, a mountain resort on the western slope of the Sierra, forty miles northeast of Bakersfield and far from the bright lights which some said he loved too well. Faced with the opportunity of a lifetime, he left no doubt he meant to take advantage of his big chance. He chopped wood and did roadwork up and down hills. He sent back word that when he came out, he would be well along toward being in the best condition of his life. Acknowledging the challenge of a clever, hard-hitting opponent like Langford, who shuffled forward flatfooted, ready to deliver a finishing blow at almost any time, he announced he would sharpen his speed by sparring with Kid Booker, the Bakersfield lightweight.

Even as Lester prepared, word reached Taft from the east that Langford, enroute by train, was not in the best condition, that his waist looked massive, that his stomach hung over his belt. There were rumors that he looked slow, that his reflexes did not seem as sharp as they had been. The talk only heightened interest. Perhaps Langford, for all his reputation, was ready to be taken.

Interest had reached such a point by the Thursday night, ten days before the match, when Langford and his party arrived by train in Bakersfield that there was a crowd estimated at more than three hundred persons waiting to see the famous boxer. Langford's entourage included his manager, Joe Woodman, his sparring partner, John Davis, and his trainer, Bob Armstrong. Armstrong had been a big-time fighter himself, having once gone ten rounds with Jim

This was the scene on a sunny October afternoon in 1913 when the local favorite, Jack Lester, challenged the number one heavyweight contender, Sam Langford, in a 20-round contest at Taft. (Photo from Howard H. Bell, Jr.)

Jeffries, whom he later trained for Jeffries' ill-fated comeback try against Jack Johnson. He also had to his credit a four-round no-decision match with George Gardner, who had been the world's first recognized light heavyweight champion. The group was a day late arriving, having run into the very problem which, by their presence, they were in the process of correcting. Like so many before them, they had not known where Taft was. They had taken the wrong train at Richmond, and it had delayed them a day. Joe Woodman's first act on getting off the train in Bakersfield was to send Langford off to bed to begin recovering from the rigors of the transcontinental train trip. His second act was to vehemently deny that Langford's weight had been 199½ pounds as reported in the press for the Jeanette fight. He said Sam had come in at his usual fighting weight, had been then and was now in the best of shape.

The following day Sam Langford and his party traveled to the West Side on the Sunset railroad. As the train passed through Maricopa, there was a large crowd on hand to see the leading contender. There was an even larger crowd, estimated at 500 persons, on hand at the station in Taft. After a friendly welcome, Langford and his

attendants transferred their baggage to the cottage Promoter Wilson matter-of-factly provided them on Center Street. After the party settled in, Joe Woodman lost no time setting up a training camp in the Mariposa pool hall, where a ring had already been constructed.

As if in recognition of the arrival in Taft of the number one heavyweight contender, nature unveiled a spectacular show. At 3:30 A.M. at the No. 21 well being drilled by Kern Trading & Oil Company on Sec. 27, 31S-23E, four miles northwest of Taft, the crew was drilling at 2,930 feet, expecting pay sand fifty feet deeper. Suddenly the well began to kick, catching the crew unprepared. A column of oil estimated at 25,000 barrels a day shot from the hole, flooding into a nearby sump. No effort was made to control the well. Men were routed out of bed to throw up dikes to contain the oil. The first light of day was just beginning to brighten the sky when a whoosh from the sump, where a pumper had been seen with a cigarette, signalled new danger. Oil had caught fire. Flames shot three hundred feet into the air, lighting the sky as if the sun had risen in one quick movement. At the nearby Mays No. 9, crews began playing water on the derrick. At the nearby Caribou No. 3, which had come in a week earlier making 3,000 barrels a day, men worked to build an earthen barricade that would stave off the seething oil from K. T. & O.'s wild well.

Not long after the well caught fire, Sam Langford was up and on the road, traversing the trails of 25 Hill in back of Taft, acknowl-

edging the friendly shouts of men at work in the oil field with a
wave and a smile, jogging along in heavy clothing that made it impossible to tell whether he was carrying excess weight. He racked
up eight miles before returning to his cottage, where he rested and
ate. In the afternoon, he was out on the road again. Later he put in
an appearance at the pool hall, where one hundred seats had been
provided for spectators. Admission was free, and the place was
filled. Langford, who wore a heavy loose-fitting sweater, engaged
in friendly banter with the fans. Then, without removing the sweater, he sparred four rounds, the first two with Bob Armstrong, the
last two with John Davis. He finished up by skipping rope and
punching the heavy bag. It was a long first day in Taft for the big
boxer.

Any lingering doubt that the city of Taft would become world-
famous through the staging of the Langford-Lester fight was dis-
pelled with the announcement that the match would be filmed by
Norbig Film Company, of Los Angeles. The company's president,
Frank Norton, said the film would be shown all over the world. The
cameraman had hardly arrived in Taft before Norton decided to
enlarge the scope of the project. Not only would the film include
round-by-round action of the fight, it would in effect be a docu-
mentary of the West Side, including pictures of prominent businessmen, their establishments and their families. It not only would show
the fight, it would also introduce to the world the people of the West
Side. As a starter, the cameraman set up his movie camera at Conley
School, catching the several hundred children who went there as
they raced out after Principal Hamilton turned them loose.

Thanks to the flaming K.T. & O. well, an aura of carnival hung
over the West Side. The nights were bright with the light of the
burning well. A carnival did, in fact, move into town to take advantage of the crowd assembling for the boxing match. It was the
McAllister & McBride Amusement Company carnival which had
played the state fair at Sacramento. It featured monkeys, a dancing
show, a merry-go-round and a Ferris wheel. The Ferris wheel afforded a spectacular view of the K.T. & O. gusher, which was
drawing a steady stream of visitors. Among them was Joe Woodman, Langford's manager, who motored out to the well site with Al
Israel, proprietor of the Crawford Bar. Woodman said it was a sight
he would not have missed for the world, adding that he felt the

people of Taft were according him and his protege "the best of treatment."

Another visitor to the well site was the Norbig Film Company cameraman, who decided to enlarge the scope of the West Side picture beyond the Langford-Lester fight and the people of the community. The film would show scenes of the burning gusher. The spectacular show attracted yet another professional film firm, Keystone Film Company, of Los Angeles. With Walter Wright in charge, the Keystone crew began shooting footage, announcing they would use it in a film, the plot of which would be worked out later. The well appeared to appreciate the arrival of the Keystone crew, the *Bakersfield Californian* reported, because it "roared and soared higher than ever, the great cloud of smoke rolling into the heavens, carrying with it the red flashing flame, thus forming a picture which is impossible to comprehend without seeing."

The burning well formed an awesome sight. A sobbing roar accompanied what a newsman described as "belching flames that make it look as if the fires of Hades have been set loose." By day, the towering column of smoke darkened the sun. By night, flames could be seen from as far off as Old River, twenty miles away. For the oil company involved and its neighbors, it was far from a welcome show. K.T. & O. officials calculated the loss at a minimum of $10,000 a day. There was the strong possibility that the fire would spread to other wells, not only K.T. & O.'s but also adjacent companies'. More than one hundred men labored day and night to contain the oil and devise a method to put out the fire. The likeliest approach, one that had worked at the Pacific Crude blowout a year earlier, was to blast out the fire with steam. K.T. & O. began lining up boilers for the effort. Another thought was to snuff out the fire with dynamite. A crew began tunneling toward the 10-inch surface casing.

As the pavilion neared completion, Promoter George Wilson announced ticket sales were brisk. There were four classes of tickets, including $10 for ringside, $7, $5 and $3 for seats farther back. Wilson said he had received orders from San Francisco, Los Angeles, Salt Lake City, Denver and many of the larger mining camps in Nevada. Al Israel of the Crawford Bar announced plans to publicize the fight by releasing paper balloons from in front of the Crawford at 2:00 P.M. on the Sunday preceding the contest. In Bakersfield, Santa Fe announced it would run a special train to Taft on the day of the fight. The train would leave Bakersfield at 11:00 A.M. after making connections with a special from Fresno, arriving in Taft in time for the four-round curtain-raiser at 1:30 P.M. Roundtrip fare was $1.50. The reason for the early starting time for the fight card was to ensure adequate light for the filming of the Langford-Lester contest, which was to begin at 2:30 P.M.

Meanwhile Keystone Film Company had worked out the plot of the feature movie it was filming at the scene of the K.T. & O. gusher, which was still aflame and showed no sign of abating. The plot would involve a beautiful young lady who had inherited a supposedly worthless property. The property would be the scene of the gusher. The villain, who had been spurned by the heroine, would creep up to the well and set it afire, after which he would flee town, pursued by irate citizens. Shooting began on the Mays property adjoining the K.T. & O. lease, with the actor Charles E. Insla stealthily creeping up to set fire to a pool of waste oil. The cameraman panned from the waste oil to the actual fire at K.T. & O.'s No. 21, giving the impression Insla had set the monumental blaze. For the chase scene, the film company hired various Taftians, among them Mr. and Mrs. A. T. Connard, Mrs. Electa Cudney, C. Albrecht and A. M. Keene, publisher of the *Driller*. The pursuit of Insla to the train station took on an added note of realism when a passerby, Dr. Kleiner, seeing his friends in hot pursuit of a man who was a stranger to him, took up the chase and narrowly missed catching the fleeing actor before he swung aboard the train.

The burning gusher was far from a frivolous matter to its owners. Seven days after the well had blown in, they were ready to make their first major attack. They had brought into position fifteen boilers. They had built galvanized iron shields to protect the crews that would man the 8-inch steam lines. They had constructed a 2,000-barrel tank and filled it with mud. Shortly before 10:00 P.M. on a night made hot by burning oil, K.T. & O.'s crews struck back, hitting the flaming oil with a massive dose of steam followed by a deluge of mud. Flames were extinguished. In the darkness, workmen poured on steam. Within minutes heat in the sides of the sump reignited the flow of oil and flames danced into the night sky.

Jack Lester, meanwhile, was planning his attack. At training quarters in Woods' Bar on Center Street, he looked good in sparring sessions with Chick Brown and Kid Ketchell, impressing observers with speed and ringmanship he had not shown before. The *Driller*, describing Lester as "fast and peppery," reported, "Lester is made of iron. He will be hard to tear down no matter how hard Langford can hit—should Langford be able to land a blow. Lester has grown more clever and the wise ones are keeping their eyes on this lad." Lester's big chance loomed larger and larger. The eyes of the sports world were on him. Kern Mutual Telephone, in confirmation of interest, was setting up special lines to the pavilion to make sure word of the outcome would be flashed to a waiting world without delay. Depending on who won, Jack Johnson might have his next opponent. The fight would probably be held in France, but there would be nothing wrong with that. It had been only twenty-four years before that John L. Sullivan had defended the crown there against the Englishman Charley Mitchell.

The shot at Johnson's title only worked one way. For it to materialize, Lester had to whip Langford. It would not work the other way. Johnson was a Negro. It was the era of the white hope—the search for a white man to beat Johnson. Langford, though credited with clean living habits and gentlemanly demeanor, had only an outside chance of getting a shot at the title, for he was black too. If he had been white, he long before would have had a chance at the championship and, according to many, would have come away in possession of it. For Lester, it was a golden opportunity. His friends quoted him as saying he would win before the 12th round. They pointed out that Langford had not trained for the 20-round distance for more than two years, that he was getting older and was overweight. It would be Lester's strategy, they said, to drag the fight out, to catch Langford when he tired and move in for the knockout. And then, off to Paris for a showdown with Jack Johnson.

Langford refrained from precise predictions, saying he intended to win when he could. He continued with strong emphasis on roadwork, knocking off at least eight miles a day. Afternoons, he sparred with Bob Armstrong and John Davis, impressing those who watched with his skill. He was adept at feinting, repeatedly drawing his sparring mates off balance, countering effectively to the body. There could be no question about one thing. He knew his business. The question was whether he could stand up to a strong young man like Lester, who was seven years his junior. On the Saturday afternoon before the fight, at his last heavy workout, Sam Langford for the first time since he had begun training at the Mariposa took off the heavy sweater he had always worn in the sparring ring. Far from sheltering a layer of fat, his waist was flat and trim.

In mid-morning of the day of the fight, a large number of Ford motor cars assembled on Main Street and paraded down Fourth Street in a column stretching for six blocks. Norbig Film Company recorded the scene. It, too, was to be included in the film with pictures of West Side businessmen and school children, K.T. & O's flaming gusher, and the round-by-round action of the Langford-Lester contest. Stebbins & Dowd, the local Ford agency, had arranged with Norbig to include the parade, then had requested all Ford owners on the West Side to assemble and "be photographed in a picture which will be seen in theaters in all parts of the country." Other motorists pulled into town from Fellows and Maricopa, Coalinga and Bakersfield until the streets of Taft were clogged with the most massive traffic jam ever seen in the oil community.

Shortly before one o'clock, the special train pulled in from Bakersfield, bringing with it an estimated one thousand boxing fans, who trooped from the train in an eager column and began making their way toward the pavilion on Fourth Street. The arrival of the train posed a potential problem, for it included a special section reserved for Negroes. Promoter Wilson, mindful of Taft's reputation as a white man's town, had foreseen the problem and taken steps to avoid it. As Negroes left the train, they were advised that if they cared to eat while in Taft, they would be welcome to do so at the local Chinese restaurant.

In honor of the big day, business establishments except for bars and restaurants were closed, and oil companies, cooperating as much as possible, slowed activities to allow a maximum number of field men to attend the boxing contest. All oil activities could not come to a halt, like the battle to tame K.T. & O.'s gusher. For the disappointed men who could not attend the fight, Al Thackery conceived a plan for spreading the outcome through the fields. At the pavilion, there would be two dirigibles of the Zeppelin type, each twelve feet long. If Langford won, only one dirigible would be released. If Lester won, two dirigibles would be released. If the decision was a draw, neither dirigible would take to the air.

Oddsmakers quoted odds of two to one favoring Langford as fans began filling the pavilion. It was a comfortably warm day and many in attendance took off coats to enjoy the contest in shirtsleeves. Most of the predominantly male audience wore hats and ties. In the $3 seats a few men in coveralls and open-neck work shirts looked as if they had barely made it in from the field. For at

The flaming K.T. & O. gusher, above, threatened to steal the scene from the Sam Langford-Jack Lester boxing contest. Some thought it a wilder well than the Pacific Crude gusher, left, which had blown in sixteen months before. (Photo above from Phil Witte. Photo left from *Daily Midway Driller*)

least some fans, there was more excitement than they had anticipated. A section of the recently completed stands collapsed, plunging one hundred men to the ground twenty feet below. Two were injured seriously enough to require hospitalization. One was an oilworker, the other an itinerant peddler who had come to town several days before selling birds.

Promptly at 1:30 P.M., Kid Kenneth, filling in as referee for preliminary matches while he waited in Taft for a proposed match with Jess Willard, called the contestants in the four-round curtain-raiser to the center of the ring. The boxers were Billy Alvarez, a favorite with Taftians, and Johnnie Deuel, a newcomer. The match ended quickly with the clever Alvarez scoring a second-round knockout. Next up was a scheduled six-rounder, the only other match before the main event. It pitted Johnnie Pryor, another Taft favorite, against Kid Hess, who had not appeared in Taft before. The bout ended almost as quickly as it began with Hess going down for the count in the first round. The crowd, thinking Hess could have tried harder, hooted him out of the ring.

The time had come for Sam Langford to meet Jack Lester in the contest that might produce the next challenger for the heavyweight championship of the world. For the city of Taft, it was an event one newsman later would describe as "the greatest in the sporting history of the oil fields." For the moment, Taft was the center of the sporting world. At ringside were boxing writers representing the sports pages of newspapers in Los Angeles, San Francisco, Fresno and Bakersfield.

Among the estimated 5,000 persons in the pavilion, there were at least two men who had no doubt about the outcome of the impending battle. One was Sam Langford. In his dressing room, he calmly prepared for the contest by lighting up a cigar. In succession, he smoked three cigars. The brand was St. Elmo, his favorite. His new-found friend Al Israel kept him well supplied, bringing fresh cigars each day. If Sam had any worries, he kept them to himself. It would be rumored later that he had made a pre-bout deal, that for a stipend of several hundred dollars he had agreed to let the match last at least six rounds for the sake of the movies to be made of the contest.

The other individual with no doubt of the outcome was Jack Lester. Young Ketchell and Chick Brown, who had served as sparring partners and were to work in his corner, would later tell friends that Jack seemed drawn and weary. They thought he had overtrained. Lester was late coming to the ring. Long minutes after Langford had climbed through the ropes, wearing a cap and bathrobe, accompanied by his seconds, Bob Armstrong, Jack Davis, Kid Fero and Joe Woodman, Jack Lester climbed into the ring, followed by his seconds, Ketchell, Brown, Pete Morrison, Monte Michaels and George Wilson.

Lester was not to be let off easy. Any thought that the sooner he could get it over with, the better, was not to be borne out. First, there were pictures to be taken, both by Norbig Film Company's cameraman and a photographer from Clendenen Studio, who took panoramic stills of the crowd. Lester's seconds had trouble getting him to face the camera for the stills. Then there were challenges, including one by Kid Kenneth, who was there in person to challenge the winner, others by telegram, including one from Jess Willard.

Kid Kenneth, a Taft favorite who challenged the winner of the Sam Langford-Jack Lester fight, was one of many well known fighters who trained at "Pop" Robinson's Bar. Left to right, K. O. Coffee, Signal Hills Billy Fluery, Frank Mandell, Billy Alvarez, Al Palzer, unidentified bystander, Kid Kenneth, "Pop" Robinson, Vic Hanson, Abe Attell, Sailor Petrosky, Windy Windsor and Shorty Staub. In the background with apron, bartender Mustachio Mickey Schultz. (Photo from *Daily Midway Driller*)

This was followed by a request from the announcer, Tommy Pettit, that in deference to ladies present, those given to profanity at exciting moments please restrain themselves during the contest. A newsman later wrote, "The request seemed to have a good effect for little that would offend the ladies was heard." It was time then for Referee George Blake, who had come up from Los Angeles at the recommendation of Promoter Tom McCarey, to call the two heavyweights to the center of the ring, give them their instructions and send them back to their corners to await the opening bell.

In the first round, the two men exchanged lefts and rights with Langford getting the better of each exchange. At the end of the round, Lester was hanging on. In round two, Langford set up his man with a left to the head, following up with a right to the body. Lester went down for a count of seven. He got to his feet, landed a light right to Langford's body, and hung on, returning to his corner groggy. In round three, Lester backpedaled. Langford pursued him, landing light rights and lefts, dancing a jig. Lester landed a right and Langford laughed. The two men exchanged blows and a right from Lester cut Langford's lip. Langford retaliated with a solid right to the body. At the end of the round, Lester was hanging on.

In round four, Langford scored with a combination, a right to the body, a left to the head, a right to the body. Lester went down for a count of four. The two men exchanged rights to the body. Lester went down for eight. The next flurry was all Langford's. A left to the body put Lester down for six. He no sooner got up than he was downed by another left to the body. Up again at a count of five, he caught a right to the chin and went down, falling flat on his back. The bell sounded, and Lester's seconds dragged him to his corner. When the bell signalled the start of the fifth round, a towel came fluttering in from Lester's corner. Referee Blake declared Langford the winner by a knockout in the fourth round. Within minutes of the end, Sam Langford and his entourage were seated in Al Israel's seven-passenger Studebaker, their bags safely packed before the contest, speeding off to catch the train, Langford for Winnipeg, Canada, where he was scheduled to meet an unknown called the Ghost; Woodman for Boston, where he hoped to close a match with Gunboat Smith, the fourth-ranked contender; John Davis and Bob Armstrong for San Francisco, where a few days later Armstrong would say of Jack Lester, "Poor fellow, he didn't stand a chance in the world. It was merely a case of standing up and taking his beating. If Langford had really tried, he might have killed Lester."

In the wake of the match, DeWitt Van Court of the *Los Angeles Times* wrote, "It was the old story of a selling plater (an inferior horse) against a stake horse, of a good game willing boy against a seasoned man."

The *Driller*, which had described Lester as "made of iron" and said "the wise ones are keeping their eyes on this lad," reported, "The contest proved two things conclusively: that Sam Langford is the greatest heavyweight in the world and Jack Lester is worse than no fighter at all. Lester was like a baby. He was just 'not there.'"

The *Bakersfield Californian* found a certain poignancy in the contest, reporting that "Lester, the obscure fighter who succeeded in bringing Langford out from the East, would have found himself in the most conspicuous position in the pugilistic world had he won." In another column, the newspaper noted that Lester was not the only loser. The unannounced gate was said to have left George

Wilson holding the bag for $2,000. Langford was said to have collected $5,000.

With the Langford-Lester contest out of the way, betting turned to K.T. & O.'s gusher. The company, continuing its efforts to snuff out the fire, repeatedly hit the blazing oil with combinations of dynamite, steam and mud. Each time workmen succeeded in stopping the fire. Each time heat from the sump reignited the blaze. Bettors placed money on the length of time, generally a matter of minutes, the fire would be out, going so far as to designate an agreed-on timekeeper so there would be no arguments in settlement of bets. On Monday night, November 3, one week after the fight and sixteen days after the well had blown in, the flames were finally extinguished for good. The company snuffed out the fire by flooding the burning well with 10,000 barrels of what the *Driller* described as "chemicals used in fire extinguishers." Oil continued to flow out of the crater, which was thirty feet across and forty feet deep. The rate was estimated at 2,000 barrels a day.

A week later, Norbig Film Company announced its film was ready for showing and that it would be premiered in Taft. The film, the announcement said, showed interesting pictures of the famous K.T. & O. gusher as well as views of Taftians, including the town's school children. Almost as if it were an afterthought, the announcement added that the film would also show rounds of the recent boxing exhibition, which, the announcement concluded, "should furnish a few minutes' entertainment."

| STORM EDITION | # Daily Midway Driller | STORM EDITION |

Volume VII   Taft, California, Thursday, January 27   Number 51

TAFT CUT OFF FROM THE WORLD

Second Storm of Month Does Over $1,000,000 Damage

Taking with it nearly two thirds of the derricks of the fields, crushing tanks like egg shells, removing roofs from houses, and collapsing many buildings, the worst storm in the history of the California Oil Fields has left the West Side in a state of idleness.

Telegraph, telephone and electric lines are down throughout the district with indications that electric power will not be furnished the field for at least another twenty-four hours.   The Western Union and Kern Mutual Telephone Company have made temporary repairs and are handling special service under difficulty.

Hundreds were made homeless by the storm with every effort being put forth to provide for their comfort.   Rev. Luther Rice of Methodist Church arranged for a number of families to make their homes temporarily in the edifice while others were taken in by friends whose homes stood the wrecking force of the wind.

Though lists of losses are far from complete the following gives a small idea of the devastation:

Field Losses	Rigs	City Losses	Accidents
General Petroleum Co.	92	West Side Bottling Works-	Weaver Pittman of the San
Standard Oil	59	Roof missing.	Joaquin Light and Power
Honolulu	27	Office Bar-$300. electric sign	Company is suffering with

| Continued on Page 3 | Continued on Page 3 | Continued on Page 3 |

MUCH DELAYED BUT STILL ALIVE

Though it is impossible to move the big press of The Daily Midway Driller plant and all machinery is tied up because of the lack of electric power, The Daily Midway Driller is far from being in the land of the idle.

This paper though small in form carries much of to-day's live news.   It was issued on a small press with the use of hand power.   In getting out yesterday's paper to-day we are somewhat late but we believe our efforts, although against great odds, will be generally appreciated.

The Great Windstorms

The weather on the West Side turned unusually cold as the year of 1915 drew to a close, dropping to 20 degrees on a late December night in the Sunset district at Maricopa. George H. Lowell, manager of Northern Water Company, which served oil companies in the field, spent the night directing men in keeping water moving through pipelines on top of the ground that carried water to keep up steam at drilling wells, where work went on 24 hours a day. As a result of the vigilance, few, if any, pipes froze hard enough to burst. Weather warmed. Men turned to the celebration of New Year's Eve with parties at the Petroleum Club and the Blaisdell Opera House in Taft as well as at the Maricopa Club, a newly opened private club which boasted a dance floor said to be among the finest in the oil fields.

The fervent hope was that the year of 1916 which was just beginning would be a better year than the one which had ended. Probably at no time during the history of the oil industry in California had the operator been confronted with as many difficulties as prevailed during 1915. The early months of the year had found an over-production of approximately 30,000 barrels a day. South American markets had been closed by the war in Europe. There had been a general contraction in business on the Pacific Coast. The federal government, in the wake of a Supreme Court decision in February, 1915, finding the withdrawal by President Taft some three years before of more than 68,000 acres of public land for the setting up of Naval Petroleum Reserves at Elk Hills and Buena Vista Hills near Taft to be valid, had initiated 25 oil land dispossession suits and was threatening others against operators on unpatented lands in the withdrawn area.

The uncertain situation had contributed to a decline in development work. Not since 1906 had the state's oil industry completed so few wells, a total of only 240. As a result, production by the end of 1915 had dropped to 236,000 barrels daily, a decline of 14,000

Snow began to fall at Reward at 5:00 A.M. on Thursday, December 30, 1915, and fell off and on all day. The following day—the last of the old year—the ground was covered with snow, an unusual occurrence in what was to prove an unusual winter. (Photo by O. C. L. Witte)

barrels a day since the beginning of the year. Total production for the year added up to 89 million barrels compared with 103 million barrels a year earlier, a falling off of 14 million barrels. Exploratory drilling had declined, with only one discovery of any promise. Belridge Oil Company had moved out six miles northwest of the discovery well with which it had opened the Belridge field four years before and found 35-gravity oil at a depth of 4,000 feet on Sec. 35, 27S-20E. Other wells had been started on adjacent sections, but the extent and value of the new field remained unproved.

Though many found the oil situation as unpredictable as the unusual winter weather that beset the West Side, others saw hope that the future would be better. There was the thought that Congress would enact in its present session the remedial legislation necessary to protect those who in good faith had given time and money to the exploration of unpatented lands. A ray of hope came on January 5, 1916, with the report from Washington that the General Land Office had found the claims of Honolulu Consolidated Oil Company to twelve quarter sections of land at Buena Vista Hills, totalling 1,920 acres, to be valid. The land had been included in the second Naval Petroleum Reserve established by President Taft. The land office found that Honolulu had acquired the land before the withdrawal order and had undertaken development of the property in good faith. The belief was expressed that the government agency's action would serve as a precedent in the determination of applications that covered over 20,000 acres in the withdrawn area. The decision in Washington brought rejoicing in the field. At Honolulu's main camp in the Buena Vista Hills, four miles northeast of Taft, the celebration grew to the point where someone tied a flatiron on the whistle cord at the boiler plant and allowed the entire field to share the enthusiasm.

In the face of increasing foreign and domestic demand, it was predicted that the price of gasoline might rise to 25¢ a gallon, bringing a corresponding rise in the price paid producers for crude oil. The price for gasoline in Los Angeles and San Francisco was 16¢ a gallon, up 4¢ since the preceding August. In Paris, the price had already risen to 75¢ a gallon, reflecting the cut-off to the Allies of the Russian supply and the closing of the Suez Canal to merchant ships.

Among larger companies, Standard Oil Company professed hope for the future, noting that demand for oil was increasing and that prices were going up. Even in the preceding year's production decline, a spokesman in the *Bulletin*, a company publication for stockholders, found a brighter side. The spokesman took the view that the decline was small considering the lack of development drilling and illustrated "the marvelous stability and permanency of the California fields." The article concluded, "It appears quite possible that the present production might be maintained for a considerable time with but little development."

Even with the decline in production, California still ranked as the nation's number one oil-producing state. The state's production of 89 million barrels during 1915 topped Oklahoma, which produced 80 million barrels, and was far ahead of other states in the top ten, including Texas, 26 million barrels; Illinois, 18.5 million barrels; Louisiana, 18.5 million barrels; West Virginia, 9 million barrels; Pennsylvania, 8.7 million barrels; Ohio, 7.9 million barrels; Wyoming, 4.2 million barrels; and Kansas, 3 million barrels.

In California, Midway-Sunset continued to produce more oil than any other district, accounting for 40 million of the 89 million barrels the state produced in 1915, or almost one out of every two barrels. The nearest competitor was Coalinga, far behind with a production of 13.5 million barrels. The Whittier-Fullerton district was in third place with almost 13 million barrels.

A forest of wooden derricks offered visible evidence of the West Side's preeminence, extending for a distance of more than 25 miles from Maricopa northwest to the Reward area near McKittrick. There were more than 2,300 derricks in the man-made forest. Most were standard cable tool derricks, which normally ran 72 to 84 feet high. Some were as tall as 130 feet, marking wells that were deeper than average. The sturdy derrick, best known of any oil field symbol, owed its name and presumably its origin to an Englishman named Derrick who, in about 1600, built the forerunners of the oil field structure to carry out his assigned task, which was hanging people.

It generally took a crew of five or six men two or more days to put up a standard derrick. The crew used a crosscut saw to cut timber to size manually. As men worked, they moved higher and higher off the ground, lifting heavy timbers by rope, scrambling about to nail each piece in place. There was no possibility for use of a safety belt. The only comfort a man might take was that he was not apt to fall more than once. The rigbuilders were an elite group, performing what was acknowledged to be the hardest and most dangerous work in the oil fields.

The role of the wooden derrick was not confined to that of support structure for equipment used in drilling the well. After the well was completed to production, the walking beam that had been used during the cable tool drilling process was converted to new service, bobbing up and down to pump the well. If the hole was drilled with rotary tools, which had been introduced on the West Side in 1908 but were not yet in widespread use, there was still a need for the derrick. When the well required remedial action, such as pulling pipe to clean out sand or replace the pump, the derrick served the same support role it had served during drilling. Each derrick represented an investment of up to $4,000. Some cost more, like those on Standard Oil Company properties, which were strongly guyed to withstand wind and set on concrete bases. The expense was considered necessary in light of the derrick's vital role not only in drilling the well but also in keeping it on production.

As January unfolded, weather on the West Side worsened. Heavy rain fell, soaking into the ground to turn oil leases into muddy quagmires. In Taft, the threat of more rain failed to dampen enthu-

Reward area of the McKittrick field, like the rest of the West Side, was a forest of wooden derricks in early January 1916. (Photo by O. C. L. Witte)

siasm for the events that were being offered. Boxing fans flocked to the Blaisdell Opera House, paying one dollar admission to watch matches promoted by L. R. Buchanan and John Pyles. The hit of the Friday night, January 14, card was Young Newsy, a newcomer who had come to Taft riding the rods and whose diet for two days had been doughnuts and coffee. Matchmaker Pyles put him in against Young Willard, who suffered a dislocated thumb in the first round. Newsy, seeing that he would have no chance to make a showing against Willard, stepped to the center of the ring and challenged any 125-pound man in the house. Terry Webster accepted the challenge. While Webster was getting ready, fans showered the ring with nickels, dimes, quarters and dollars. Webster and Newsy put up a fast first round. The second saw Newsy holding his own for perhaps a minute when there was a noise of snapping bones, and the boxer dropped his right arm to his side. Dr. J. Walter Key, who was quickly called, found a bone had been fractured.

While the disappointed boxer was leaving the ring to the plaudits of the crowd, there was also a good attendance at the Presbyterian Church's Friday night lecture, which was titled "Evidence for the Deity of Jesus in the History of the Church." The lecture was one in a series the church was sponsoring. Another event attracting attention was the clearance sale in progress at Smith Brothers, which featured Hart Schaffner & Marx suits reduced in price from $28 to

$21, Stetson hats reduced from $15 to $12, and a five percent discount on all shoes.

On Monday morning, January 17, men went to work as usual in the oil fields. A light rain during the night had swelled the season's total on the West Side to more than three inches, which was above normal. Rain had stopped, and the day dawned under a clouded sky. In the oil fields, men proceeded with the work of drilling new wells and keeping old ones on production, laying lines and gauging tanks. Shortly after nine o'clock, wind swept over the Temblors, blowing from the southwest. On the San Francisco-McKittrick Oil Company lease at Reward, the wind blew the crown block from a derrick. The block smashed into a boilerhouse, causing the boiler to explode.

A pipefitter unhitched his horse near the store at Reward and told the animal to go to the barn. Minutes later the roof blew off Pacific Oil Company's office, landing on the pipefitter's wagon. The horse arrived at the barn unscratched.

Well gangs began unhitching their horses, sending them to the barn. Even as horses ran a gauntlet of flying particles of timber, tar paper and sheet iron roofing, following the winding road that led through a short steep canyon to the barn, men hurried to their homes, fearful for the safety of their families.

An Associated Oil Company crew hurried to tighten guy wires on derricks near Station H. They had completed the work on three derricks when the wind, blowing with increasing fury, toppled all three derricks.

At Associated's Station H, three employees began shutting off fire from the boilers. Finishing, they hurried from the building just as the wind blew the roof off. A falling door struck one man, hurling him against a boiler. He ducked under the door and succeeded in reaching safety.

Smokestacks on Stations N, J and L began to sway. The wind tore them from their supports, shutting down operations.

Everywhere wooden derricks were falling. Roofs flew off buildings. Wind crushed tanks. On the Kern River Oil Company lease, a bull wheel weighing 2,000 pounds rolled down the steep hill at the rear of the residence of Superintendent H. G. Ball. Fortunately the wheel stopped twenty feet short of the building.

On Associated's lease near Station N, the William Wheeler family, fearing for their safety when wind tore away a portion of their

Fallen derrick and damaged lease houses, Reward, January 1916. (Photo by O. C. L. Witte)

Downed derricks, shattered trestle, State Consolidated Oil Company, Reward, January 1916. (Photo by O. C. L. Witte)

Collapsed tank spewed pool of oil, Reward, January 1916. (Photo by O. C. L. Witte)

house, went to the residence of a neighbor whose house was sheltered. Wind tore the Wheeler house from its foundation, moving the structure a distance of fifty feet. The house survived intact except for the loss of its roof. When Wheeler entered the house later, he found nothing in it had been greatly disturbed. A clock still ticked steadily on the mantle.

By noon, wind blew with hurricane force, bending smokestacks on Berry & Keller Oil Company's station over the building, stopping use of the plant. Wind leveled J. D. Scott's garage and overturned his car. It damaged the office and a bunkhouse on Kern Trading & Oil Company's lease as well as the residence of the superintendent, S. L. Fry. On the Del Monte lease, wind damaged the residences of E. O. Peters and W. E. West.

Even as high winds tore at McKittrick and Reward, winds of less intensity swept North Midway and the Twenty-Five Hill area in back of Taft, felling derricks, crushing tanks and tearing the roofs off buildings.

On Twenty-Five Hill, a team being driven by Paul Garretson, a gauger for Standard Oil Company, became frightened by flying timbers and bolted. At the Mascot lease, the runaway team became tangled in debris. Garretson had to cut the team loose from the wagon to free the animals.

Windstorm of January 17, 1916 leveled derricks at Reward. (Photo by O. C. L. Witte)

Wind tore away the back of the Star Theater in Fellows during the January 27, 1916 storm. (Photo from Phil Witte)

Elsewhere on the Mascot, Mrs. J. N. Ripple, whose husband was the superintendent, saw a telephone pole go down, badly twisting the wires. She restrung the wire, restoring service.

In early afternoon some three hours after the storm began, wind died down and the sun came out, leaving oil men to survey a scene of devastation. A rumor swept the fields that a man had been blown out of a derrick and killed. The rumor proved untrue. The oil-worker was Ernest Stroshine, who had been working up a derrick in the McKittrick area when the storm struck. Hurrying to come down, he had lost his footing near the ground and fallen, fracturing his right ankle. Fellow workmen had rushed him to General Hospital in Taft, where Dr. Page had made an X-ray and confined him to bed.

Wind had barely subsided before Phil Bowles, superintendent of Reward Oil Company, revved up his Duesenberg, drove past the flattened rubble of downed derricks and sped into Bakersfield to buy all the lumber he could get. Before the storm, there had been thirty-nine derricks on the Reward property. Afterward, thirty-four lay wrecked.

Oil field superintendents rushed for lumber yards. In Taft, Jack Bennett of Section Twenty-Five Oil Company, which had lost six of its thirty-five derricks, was the first into the Moron Lumber Company yard. He contracted for lumber for six derricks. J. N. Ripple of Mascot Oil Company followed with an order for lumber for four derricks for immediate delivery and enough for six more for delivery in the next ten days. Traders Oil Company entered an order for lumber for five derricks, Wilbert Oil Company an order for lumber for four. Don Knoles, manager of the yard, telephoned an order to his suppliers for 200,000 feet of lumber for delivery within a week.

At Union Lumber Company in Taft, Manager F. E. Davis signed contracts with General Petroleum Company, Pierpont Oil Company, Hale-McLeod Oil Company and California Counties Oil Company for immediate delivery of lumber for derricks and made arrangements to supply others as soon as shipments could be brought in. Davis said the lumber yard would be kept open twenty-four hours a day until further notice.

In McKittrick, within a matter of hours, companies ordered one hundred carloads of lumber.

There was a rush on supply houses in Taft, where demand for rig irons, heavy spikes, belts and other equipment necessary to return wells to production equalled that of the busiest boom days.

The telegraph office was kept busy sending telegrams ordering material for the rebuilding job. Messages asked that orders be shipped express.

The storm left the community of McKittrick without water service. When derricks toppled over water wells in Little Santa Maria Valley, timbers smashed surface connections. Associated Oil Company, which drew water from the same wells, rushed in a crew to repair the damage and rebuild the derricks.

Wind downed San Joaquin Light & Power Corporation lines that carried electricity to the North Midway district, leaving the area without service for hours.

Western Water Company, which served the community of Taft, escaped with loss of only the cupola which had been built on top of its office as a matter of design.

The count of downed derricks showed the McKittrick field hardest hit, followed by North Midway and Twenty-Five Hill, with the Maricopa area suffering little damage. Losses were estimated at $500,000 to $1,000,000.

At McKittrick, 196 of the 315 derricks that had stood prior to the windstorm had been destroyed, or almost two out of every three.

Associated Oil Company, the biggest operator in the field, suffered the greatest loss, losing 66 of 114 derricks. The second largest company, Kern Trading & Oil, lost 32 of 48 derricks. Berry & Keller lost 12 of 15 derricks; East Puente Oil Company, 4 of 17; Fairfield Oil Company, 5 of 7; Jackson Oil Company, 4 of 7; Jewett Oil Company, 9 of 12; Kern River Oil Company, 3 of 4; Midway Royal Petroleum Company, 4 of 10; San Francisco-McKittrick, 13 of 18. Hardest hit of smaller companies was Olig Crude, which had four derricks and lost them all.

When Otto C. L. Witte, a young pumper for State Consolidated Oil Company, sat down that night to write in his "A Line A Day" diary, he wrote:

"Pumped. Rainy & stormy, turning into a hurricane at noon. Blew over hundreds of derricks in the oil fields. State O. Co. lost four rigs and damaged cook & bunk houses. Worked in plant p.m."

Among those who suffered losses at North Midway and Twenty-Five Hill were American Oilfields Company, which lost 10 of 64 derricks; Buena Fe Petroleum Company, 22 of 53; Chanslor-Canfield Midway Oil Company, 32 of 115; General Petroleum Com-

Hurricane winds of January 27, 1916 storm toppled two derricks, left and right, on Twenty-Five Hill and spared one in the middle. (Photo from Phil Witte)

Incredibly, one derrick on Mascot Oil Company's lease on Twenty-Five Hill toppled on its side with no structural damage during the windstorm of January 27, 1916. (Photo from Phil Witte)

pany, 40 of 135; Indian & Colonial Development Company, 11 of 25; Mascot Oil Company, 10 of 41; North American Oil Consolidated, 25 of 73; and Traders Oil Company, 8 of 12.

Few companies had purchased windstorm insurance for derricks, though there were exceptions. Less than two weeks before the windstorm, L. P. Guiberson had bought policies through the Heath Agency covering ten derricks. Seven of the ten were destroyed.

A hurry-up call went out for rigbuilders. Men flocked to the fields from as far away as Texas, responding to the offer of a $10 a day wage to rebuild derricks lost to the storm. So great was the demand that it extended to any man who had once used a hammer or saw, though he was far from being a carpenter.

Troy Owens, superintendent of Hale-McLeod Oil Company, was the first to get a well on the beam after the derrick had been wrecked. Forty-eight hours after the derrick of the No. 1 well on Sec. 8, 32S-23E, three miles northwest of Taft, had been turned into kindling, a new derrick was in place, the beam bobbing. Potter Oil Company on Sec. 15, 31S-22E at North Midway was not far behind. The company contracted for the services of Stitzinger & Seeley, rigbuilders, who put a double crew on the job, erecting the first derrick in 36 hours. Charles Wakefield, superintendent for Pierpont Oil Company, claimed to be the first to get a derrick up on Twenty-Five Hill, though he was not the first to get in an order for lumber. Three days after the windstorm, the beam was bobbing on Pierpont's No. 3 well.

There was a call, too, for teams to haul lumber over lease roads made impassable to trucks by rain.

Motion picture companies in Los Angeles sent representatives to the West Side to film the oil field devastation. Ed Shortridge, a veteran stage driver, took cameramen on a day-long tour so they might take pictures of roofless houses, collapsed tanks and derricks in all stages of destruction.

By Friday, four days after the windstorm, sunshine and clear skies revived spirits. At North Midway, General Petroleum pointed with pride at the speed made in rebuilding the derrick over the No. 21 well on the Globe property, which had been placed on the beam hours ahead of others. On the Nevada-Midway property, the company had placed wells on a jackline system before the storm, drawing power to pump the wells from a central plant. The storm toppled fourteen out of forty-seven derricks. Within forty-eight hours, workmen had cleared away debris, repaired damaged jacks and returned all of the wells to production.

The short supply of lumber appeared ended. Sixteen carloads arrived in one train at the same time seven cars were standing on sidings for Union and Moron Lumber Companies. Men looking for work were warned from the fields as the supply of available men overtook demand.

Life returned to a semblance of normalcy. In Taft, those interested in the formation of a civilian rifle club to be affiliated with the National Rifle Association held an organizational meeting in the office of Attorney H. H. Bell in the Key Building. About twenty men were present, all signing to become members. O. E. Liddell and Dr. J. Walter Key, who had taken the lead, explained the value of rifle practice and at the same time impressed on all assembled that though they became members of the rifle club they in no way would tie themselves up for military service with the government in case of war. However, they added, if any member desired military service, the fact he had been a member of the association would place him in a position to become a member of a sharpshooters' squad. The government was to issue each member a Krag Jorgensen magazine rifle of the model of 1898 with ball cartridges to be issued annually, not exceeding 120 rounds per year.

At the Presbyterian Church's vesper service on Sunday, the Reverend E. P. Shier, pastor, spoke on the theme "Twice Born Men," giving examples of the power of the Gospel to change men's lives. One example was of "The Puncher," an ex-prizefighter in the slums of London, a man said to be so rough and hard and vicious that it seemed as though there could be no possibility of his redemption, and yet when he received Jesus, he became as a little child. Another case was one the Reverend Shier said had come under his observation in a Chicago mission. The man, who until forty years of age was a gambler and drunkard, was down and out. While partially under the influence of liquor but in deep despair, he had finally sought salvation and had become a new man.

In the face of good weather on the West Side, continuing bad weather on the coast seemed remote. A gale with winds reported reaching 94 miles an hour swept the north Pacific coast, raising fears for the safety of the 100 persons aboard the steamship *Admiral Senley*, which was enroute from Seattle to San Francisco. The coastwise steam schooner *Centralia*, enroute from Grays Harbor to San

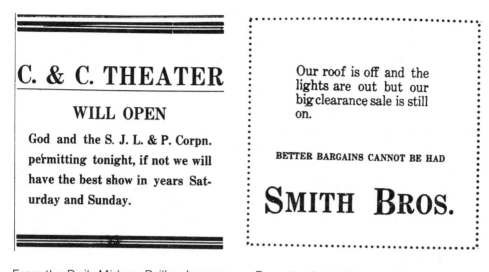

C. & C. THEATER

WILL OPEN

God and the S. J. L. & P. Corpn. permitting tonight, if not we will have the best show in years Saturday and Sunday.

Our roof is off and the lights are out but our big clearance sale is still on.

BETTER BARGAINS CANNOT BE HAD

SMITH BROS.

From the *Daily Midway Driller*, January 28, 1916. (Beale Memorial Library, Bakersfield)

From the *Daily Midway Driller*, January 27, 1916. (Beale Memorial Library, Bakersfield)

Francisco with four passengers, lost a deckload of lumber and was taken under tow by the *Governor* of the Pacific Coast Steamship Line. The tanker *Frank H. Buck*, owned by Associated Oil Company, hove to off Coos Bay, Oregon, with damaged steering gear. The storm at sea seemed far away as work of rebuilding from the windstorm of January 17 continued in the oil fields.

A light but steady rain fell over the West Side on the night of Wednesday, January 26, boosting the season's total past the four-inch mark. In the morning, rain stopped and the sun came out. Thursday, January 27, was off to a good start. Before noon, skies darkened. The wind came up, sweeping in from the west with increasing force.

In the McKittrick field, Kern River Oil Company, which had lost three of its four derricks ten days earlier, lost its fourth derrick. McKittrick Oil Company, which had lost two of four derricks, lost the remaining two. Petroleum Producers Syndicate, whose lone derrick had survived the earlier storm, became a victim. Wind drove the derrick to the ground.

Unlike the first windstorm which had struck with greatest intensity in the McKittrick and North Midway sectors, the second hurled its deadliest fury farther south. The towns of Fellows, Taft and Maricopa, largely spared before, caught the full force of the second windstorm, which would be described variously as a gale, tornado or hurricane, depending on the part of the United States the person offering the description happened to come from.

In Fellows, Otto Kramer was standing on a wooden sidewalk. The wind picked up the sidewalk and flailed him with it, breaking his right leg. Wind tore at the Star Theater, demolishing the back of the building, which was one of the largest in the North Midway. The roof flew off Smith's pool hall, shattering the glass windows of Lawton & Blanck's general store. Other buildings were heavily damaged.

Near Taft, Weaver Pittman, a lineman for San Joaquin Light & Power Corporation, answered a call reporting a damaged line. He began to climb the pole to repair the problem. A piece of flying corrugated iron, torn from a roof, pinned him to a climbing spike, breaking several of his ribs.

Al Weaver and Bob Longtine of Oil Well Supply Company were driving through the oil fields when falling lines snagged the car, bringing their trip to an abrupt halt. Both suffered lacerations.

On the Midway Gasoline Company property, Frank McIntyre sought shelter in a lease building. Wind blew down the building. McIntyre received severe cuts about the head.

A rumor spread that an oilworker had been killed on the Potter Oil Company property.

As the full force of what would later be called sixty-mile-an-hour winds hit the West Side, telegraph, telephone and electric lines went down, blacking out power, cutting off the community from the outside world.

In the office of the *Daily Midway Driller*, Editor A. M. Keene found himself without power to put out a newspaper. With the main press immobilized, he turned to a small job press. Jack Hale and Gus Fillmore took turns pumping the hand press, and Keene began the job of putting out a tiny four-page "storm edition." He printed 200 extra copies, which were quickly sold out.

On the street, bits of roofing filled the air. Debris included the roof from the Smith Brothers store on Center Street, where the clothiers' annual clearance sale was in progress. Advertising Manager Fox advised one and all to hurry and select bargains because the lid was off. The lights went out; the sale went on.

Wind lifted California Natural Gas Company's garage from its foundation, leaving automobiles untouched, and carried the structure more than a hundred feet.

The Allison & Berry garage was wrecked. Western Pipe & Steel Company's warehouse was partially unroofed. Midway Transportation & Warehouse Company's warehouse was almost entirely demolished. Wind blew the roof off Taft High School.

At the Office Bar, which only days before had been the scene of an appreciative banquet given Taft's volunteer firemen by Fred Kiester, wind destroyed a brand new $300 electric sign.

The gale blew the roof off the West Side Bottling Works, wrecked the garage at Pioneer Market, tore away part of the roof over a warehouse at Union Oil Company's retail department, damaged the roof of the C & C Theater and blew away part of the roof at Taft Garage.

On Supply Row, wind caved in the building that housed Los Angeles Manufacturing Company and tore away the roof from Lucey Manufacturing. Tanks and a warehouse at Midway Purchasing Company were damaged.

Residences were heavily damaged, roofs blown away, houses moved off foundations. As the gale ripped the roof off the home of Mrs. Fanny Staupe, a gas stove tipped over. Miss Floy Staupe coolly righted the stove, turned off the gas and hurried from the disintegrating house.

In Maricopa, almost every business house suffered damage. Wind tore away the entire front of Spahr's grocery store.

Lease houses were hard hit. At Honolulu Consolidated Oil Company's main camp in the Buena Vista Hills, building after building was demolished.

In the oil fields, high winds completed the wreckage of ten days before in the McKittrick field and inflicted heavy damage in the Midway-Sunset district. At Midway-Sunset, Standard Oil Company, whose 156 wooden derricks had weathered the first storm without a single loss, lost 69 derricks, among them some of the best in the field. The company's losses were not confined to derricks. Wind blew in the roofs of all of the firm's empty 55,000-barrel tanks in the Midway field and caved in the sides of many tanks. Roofs on two bunkhouses on Section 26 were blown off, and the roofs of two buildings at the company's pump station near Taft were torn off.

General Petroleum Company, which had lost 40 of its 135 derricks to the first storm, lost 59 more in the second, bringing losses to 99 derricks, with only 40 left standing. Honolulu Consolidated, which had escaped the first storm unscathed, lost 27 of 38 derricks. K.T. &

O., heavily hit in the McKittrick field during the first storm but spared in the Taft area, lost 37 of 156 derricks in the Midway-Sunset district. Chanslor-Canfield Midway Oil, which had lost 32 of 115 derricks to the first storm, lost 23 more to the second. Mascot Oil Company, which had lost 10 of 41 derricks on Twenty-Five Hill during the first storm, lost 24 more, leaving only seven derricks standing. Incredibly, one derrick fell intact on its side with no structural damage.

For many independent companies, though the number of derricks lost was small compared with the number lost by larger companies, the loss was total. Some which had survived the first windstorm without a single loss were wiped out by the second storm. Among those who lost all the derricks they had were Maricopa Star Oil Company, eight; Princeton Oil Company, five; Pacific Crude Oil Company and W. J. Schultz, four each; Miocene Oil Company, Midway Union Oil Company and Southern California Gas Company, three derricks each; Johnson Oil Company, Panama Oil Company and Petroleum Syndicate Oil Company, two derricks each; D. L. & W. Company, Kyle Davies, Maricopa Northern Oil Company, Maricopa Producers Oil Company, Maxwell Oil Company, Olig Crude Oil Company, Obispo Oil Company, Record Oil Company, S.K.D. Oil Company, Son & Bockman and Trojan Oil Company, one derrick each.

The loss was cruel for Olig Crude. The company had had four derricks at McKittrick. All had toppled in the first storm. The company had had one derrick in the Midway-Sunset district. The second storm claimed it.

Other independents in the Midway-Sunset district who had lost only some derricks to the first storm lost the surviving derricks to the second. Among them were Dunlop Oil company, which lost two of eight derricks to the first storm, six to the second; Knob Hill Oil Company, which lost two out of three derricks initially and the remaining derrick in the second storm; Le Blanc Oil Company, which had had two derricks and lost one in the first storm, the other in the second; Midway Peerless Oil Company, which had nine derricks and lost four to the first storm, five to the second; Paraffine Oil Company, which had six derricks and lost three to each storm; Pierpont Oil Company, which lost four of thirteen derricks in the first storm and nine in the second; and Traffic Oil Company, which had two derricks and lost one to each storm.

At Reward, when State Consolidated Oil Company's pumper Otto C. L. Witte sat down with his "A Line A Day" diary, he wrote under the date January 27, 1916:

"Thursday—Pumped. Storm & rain in p.m. Doing immense damage. Five more derricks turned over for State. Plant shut down."

On the slope of Buena Vista Hills, St. Helens Petroleum Company, Limited, earlier in January had been hailed as one of the luckiest companies in the field, success following success in the completion of eleven wells on Sec. 16, 32S-24E. The company's latest, No. 11, had come in flowing 1,500 barrels a day, representing one of the biggest wells in the area. When wind struck, the company lost every derrick, as well as a 55,000-barrel tank whose side was caved in. In adversity, luck did not completely desert St. Helens Petroleum. Superintendent Earl Bailey was directing the drilling of the No. 12 well when the gale struck. Fortunately, the crew had landed the water string, cementing off water. Drill pipe was standing in the derrick; both pipe and derrick went down. The well suffered no damage.

On the Edmonds Midway Oil Company property on Sec. 32, 32S-24E, four miles southeast of Taft, Standard Oil Company on the day before the windstorm had begun preparations to run crude oil from a 55,000-barrel storage tank into its line after purchasing the oil from Edmonds Midway. There was trouble with a pump, and the job was postponed. When wind hit, the tank, because it was full, escaped damage. C. S. Stuart of Edmonds Midway Oil had recently equipped one of four wells on the lease with a new 45-horsepower gas engine. When he hurried to the lease to inspect damage, he found the derrick over the well had fallen in such a way that the engine was not damaged.

On Honolulu Consolidated's Sec. 8, 32S-24E, property, wind created a spectacle beyond flying timbers and falling derricks. When the derrick toppled over the No. 5 well, which was one of the biggest gassers in the field, timbers broke the gas line, allowing a geyser of gas, sand and condensate to blow high into the sky.

At Northern Exploration Company's No. 1 well, the falling derrick ruptured the wellhead, creating a gusher.

At the Wilbert lease on Twenty-Five Hill, wind picked up a bunkhouse occupied by G. C. Anson and Pete Grant and dumped it into a sump hole. Clothing and effects of both men floated on the oil, damaged beyond further use.

On the Section Twenty-Five Oil Company lease, a bunkhouse was blown into a 10,000-barrel tank.

Wind moved the two-room, nearly 100-foot-long Cresceus school-house on Twenty-Five Hill off its foundation.

Even as preliminary estimates indicated that nearly two-thirds of the derricks in the Midway-Sunset district might be down, there were predictions it would take at least three years for the fields to recover and that an immediate drop in production of 20,000 to 30,000 barrels a day could be expected.

In the *Driller* office, still without power, Editor Keene published another small four-page paper in the same format as the one published with first news of the storm. He had an unexpected helper. C. G. Noble, hearing that the paper was being published by hand, could not resist the urge to offer his services. It turned out that Noble, who was city recorder, notary public and secretary of the Chamber of Commerce, had been a hand compositor and wanted to see the type come up one more time.

Keene called the newspaper of Friday, January 28 the "Oil Spirit Edition." The headlines read, "Oil Spirit Arising From Debris. Midway-Sunset Fields Reviving From Effects of Storm."

The newspaper reported, "Taft and the West Side after being thrown into a state of unequaled disinterest brought about by the devastation of many homes and more than a million dollars worth of lease chattels, is slowly recovering with here and there about the fields signs of activity becoming apparent as superintendents are overcoming the debris which is cluttering properties from Pentland to Lost Hills.

"Those compiling lists of rigs down and damage done, report that the derrick losses will reach more than 700 in the Midway-Sunset district alone with McKittrick, Belridge and Lost Hills adding many more to the totals.

"The more than a hundred people made homeless by the wind which swept down their houses and carried clothing and valuables along with it far beyond any finding, are now being cared for by friends and by the Churches. Special arrangements have been made by Rev. Luther A. Rice of the Methodist Church for the care of several families in the social room of the edifice.

"At the lumber yards practically every stick of timber has been bought up with the demand now taking in all dimension material in the yards of the state and reaching as far away as Puget Sound."

Operators hurried to erect new derricks to replace those felled in windstorms of January 1916. Above, workmen on derrick on Reward lease of State Consolidated Oil Company. (Photo by O. C. L. Witte)

The newspaper reported that the rumor of a man's having been killed in the North Midway was false. Robert Abeles of Potter Oil Company gave the lie to his demise by turning up in the care of Dr. Page with a badly cut scalp and injuries about the shoulders.

While Superintendent Jack Carter of San Joaquin Light & Power directed the crews laboring to restore power, Standard Oil Company, with more foresight than most, purchased every lamp in Taft's stores, picking up twenty-seven as a hedge against the time when lights would come on. A run on candles soon exhausted the supply.

Though electric service was out, some businesses still had gas service. The *Bakersfield Californian* reported, "Were it not for the gas supplied to many of the business houses, Egyptian darkness would prevail throughout the city."

On Saturday, January 29, two days after the second windstorm, the West Side braced for another storm. From headquarters in San Francisco, Standard Oil Company warned its office in Taft that a 90-mile-per-hour gale was raging along the coast and that the company's men in the West Side oil fields should make all arrangements to be prepared should the storm reach the fields. The report put the entire city and fields on edge for hours. When the storm failed to appear, grateful men went back to work repairing damaged homes and derricks.

By Sunday afternoon, power had been restored in Taft. Soon afterward, lines were cleared and power restored to Fellows and Maricopa.

A tally of the number of derricks blown down by the two windstorms showed that 1,223 out of a pre-storm 2,302 derricks had been destroyed, or better than one out of two.

In the Midway-Sunset district, 945 out of 1,987 derricks, or almost one out of two, had been lost. Most went down in the second storm, which claimed 665 derricks. Two hundred and eighty fell during the first storm.

In the McKittrick field, the two windstorms claimed 278 of 315 derricks, or almost nine out of ten derricks, including 196 that went down in the first storm and 82 lost in the second.

Reconstruction began immediately. Standard Oil Company provided solid encouragement for the job by advancing the price it would pay for crude oil five cents a barrel, which meant the company would now pay from 48 cents a barrel for 14 to 17.9-degrees gravity oil up to 60 cents a barrel for 29 to 29.9-gravity oil. The company also announced it was raising the price of gasoline one cent a gallon, which brought the price of Red Crown, the company brand, to 17 cents a gallon.

In the first four days after the second windstorm, King Lumber Company took delivery in its Maricopa yard of 350,000 feet of lumber. Crews under the direction of C. L. Hutchison worked around the clock loading the lumber on trucks and wagons for transport to leases for the rebuilding of derricks.

Within a 10-day period, 235 carloads of lumber arrived in the stricken oil fields. Lumber companies in the San Joaquin Valley made preparations to open the logging season about May 1 for what promised to be the busiest year in history.

Experienced rigbuilders directed crews of carpenters and laborers in building new derricks. The trained builders measured materials for cutting and supervised the work of less experienced men.

On February 2, the storm claimed its first fatality. Jesse Stephens, employed by a Fellows rigbuilding contractor, was killed instantly when, while clearing debris beside a wrecked derrick on the Santa Fe property on Sec. 21, 31S-22E, he picked up a guy wire, not knowing it had fallen over a live electric line. The rigbuilder left a wife and one child residing in Fellows.

Four days later, another rigbuilder, Claud Hecker, an employee of St. Helens Petroleum Company, was working one hundred feet in the air building the derrick that was to be used to continue the work of drilling St. Helens' No. 12 well. Timbers were being raised to the plank where Hecker and Charles Hall were working. Lines became tangled. In attempting to remove the tangle, Hecker lost his balance and fell. He had come to Taft nine days before from Santa Maria to seek work as a rigbuilder. The coroner shipped his body to his home in Mercer County, Pennsylvania.

In Taft, Mrs. H. A. Hopkins, president of the Women's Improvement Club, organized a committee consisting of representatives from the various lodges and other West Side groups to proceed with plans for a grand charity ball at the Blaisdell Opera House on Washington's Birthday to raise funds for those whose homes and belongings had been damaged or destroyed.

In State Consolidated Oil Company's bunkhouse at Reward, Pumper Witte made the following entries in his diary:

1/28/16—Friday. Packed valves and got plant going by noon. Run compressor and two wells on beam.

1/30/16—Saturday. Pumped.

1/31/16—Sunday. Pumped. Pumping rods on No. 12 parted. No derrick on same so will be dead. Jackplant started p.m. with four wells.

1/31/16 through 2/9/16. Pumped.

2/10/16—Thursday. Pumped. Derrick over No. 9 well completed. No. 10 at Bunk Houses being on the beam again.

2/11/16—Friday. Pumped.

2/12/16—Saturday. Pumped. No. 9 put to pumping by jack.

2/13/16—Sunday. Pumped. Rig builders finished No. 5 and started to work on No. 12.

2/14/16—Monday. Pumped.

2/15/16—Tuesday. Pumped. No. 5 started pumping again. Rig builders finished No. 12 derrick. They went over to Shale to build one.

2/16/16—Wednesday. Pumped. No. 12 was put on the beam. Attend R. M. Lodge. (Editor's note: Red Men's Lodge was a fraternal order which met in McKittrick.)

2/17/16—Thursday. Pumped.

2/18/16—Friday. Pumped. Rec. check for Jan. work. Amt. $93. Weather quite warm. (Editor's note: The check represented payment in the amount of $3 per day for 31 days of 12-hours-a-day work.)

2/19/16 through 2/23/16. Pumped.

2/24/16—Thursday. Pumped. Rig builder returned and started to rebuild derrick on No. 1 well.

2/25/16—Friday. Pumped.

2/26/16—Saturday. Pumped. Rig on No. 1 finished. No. 13 next.

2/27/16 through 2/28/16. Pumped.

2/29/16—Tuesday. Pumped. No. 13 derrick was finished. No. 3 well's rig next. Made a shelf for the oils used on air compressor.

3/1/16 through 3/9/16. Pumped.

3/10/16—Friday. Pumped. Rig builders finished job of building derricks on lease.

When January production figures for West Side fields were compiled, they showed a drop of approximately 10,100 barrels a day from December, including 7,700 b/d lost in the Midway-Sunset, or from an average of 105,791 b/d in December to 98,094 b/d in January; and 2,400 b/d in the McKittrick field, or from an average of 8,606 b/d in December to 6,230 b/d in January.

By March, production at Midway-Sunset and in the McKittrick field had been restored to the December level. In April, operators pushed output 2,300 b/d over the pre-storm figure, producing an average of 107,381 b/d in the Midway-Sunset for a gain of about 1,600 b/d and an average of 9,340 b/d in the McKittrick field for an increase of about 700 b/d.

Polly of the Midway-Sunset

Miss Polly Perley was a young schoolteacher with "a figure still rounded, in spite of the harassing duties that have worn many a schoolteacher to the extreme of angularity." Her brown hair was tinged "with just the slightest suspicion of red" and could be coaxed "into the severest knot lowdown near the neck, or caressed into a bewitching stylish coil high upon the head." She had a firm chin, a "spirited" nose and a mouth "shaping itself into a veritable Cupid's bow." Her mischievous eyes viewed the world "with courage, but alas! not always with prudence."

Robert Downing was a cashier in the town's leading bank. Of chivalrous heart, he was "tall and broad-shouldered, with limbs so disproportionately long that they would have given him an awkward appearance had it not been for the military ease and grace with which he carried himself." He was a bachelor and man of strong opinions, among which was the belief that it had been a mistake to give women the right to vote. He was attracted to Polly Perley from the moment they first met, which occurred on a Saturday afternoon in spring at the local school. Miss Perley, though not noticeably responsive, did not immediately discourage him. Unfortunately, the course of true love was not to be smooth.

Polly Perley and Robert Downing were fictitious characters who made their appearance in 1917 in a novel set in a prosperous town in a California oil field. While they were fictitious, the unnamed school where Miss Perley taught sounded suspiciously like Taft's Conley School. The school in the novel, like Conley, occupied a new building "on the outskirts of the city" and included a "spacious assembly hall." The town itself, called Petroleum City in the novel, was a new town, only three years old at the time the story was set, "but with energy characteristic of people of the Golden State, its inhabitants early decreed that waterworks, electric lights, a women's club and last but not least, substantial and attractive school buildings were absolute necessities to their wide-awake community." The "energetic and progressive" town bore an amazing likeness to Taft.

Miss Polly Perley was a fictitious schoolteacher who taught at an unnamed school that sounded suspiciously like Taft's Conley School. In real life, Ann W. Knapp taught sixth grade at Conley when this picture was taken in December 1916. (Photo from *Daily Midway Driller*)

None of which should have been surprising, for the title of the novel in which Miss Polly Perley and Robert Downing appeared was *Polly of the Midway-Sunset*. The author was Janie Chase Michaels. The book was published in April 1917 by Harr Wagner Publishing Company, San Francisco. It sold for $1.25. It featured cover design and page decorations of a lease bungalow, wooden derricks and stock tanks by Mrs. Bertha Wenzlaff Jones and a frontispiece of two pictures of the Midway-Sunset oil field donated through the courtesy of Alice K. Tupman. The author dedicated the book "to the workers of the oilfields whose toils release the black gold of the desert and make possible a newer and better age of commerce."

Miss Michaels, like Miss Perley, was a schoolteacher. She was a small, sharp-featured woman with graying hair who taught history to high school classes at Conley School and kept a discreet silence about her age. Though dainty in appearance, she maintained good discipline in her classroom, tolerating no nonsense. Her speech reflected the accent of her native Boston.

Though for many pupils attendance at Conley School involved a daily walk past Whiskey Row in what was known as Boust City on

Life for the single man in the Midway-Sunset field might have had its lonely moments, but there was nothing to prevent a man from having daydreams. Above, a room in a Kerto bunkhouse circa 1912. (Photo from Petroleum Production Pioneers Collection, Long Beach Public Library)

the south side of the railroad tracks, past saloons and frame houses in front of which kimono-clad women could be seen, there was no mistaking the West Side's insistence on first-rate schools. Conley in the period in which Miss Michaels' story was set housed eight grades and high school classes as well. The school had been built almost as soon as the town of Taft sprang up at a then-substantial cost of $50,000. Even as the novel *Polly of the Midway-Sunset* was being published, the town was adding a brand-new $60,000 high school and school trustees were making preparations to run an auto bus through the adjoining Buena Vista Hills to pick up some 40 children desirous of attending Taft's schools from such scattered leases as the Honolulu, Petroleum Midway, Standard Derby, Midway Gas, Kern Trading & Oil and St. Helens.

L. E. Chenoweth, the superintendent of schools of Kern County, praised West Side support for education, stating that the oil field schools were "giving the many pupils every facility that they may

Tennis at Conley School, circa 1914. (Photo from Kern County Museum)

advance as rapidly and with as great an amount of learning as they would if they attended the city institutions." Attendance was on the increase, rising to a new high of 926 in 1917, compared with 681 the year before.

For its part, the *Daily Midway Driller* offered every support to the local school system, regularly publishing each honor roll and offering front page space not only for commencements but also for such things as the Taft debating team's contests with Bakersfield and Delano schools. On the question "Resolved: The U.S. government should own and operate all telephone and telegraph systems in the country," Taft's affirmative team won in a three-cornered competition.

Along with schools, the town was proud of its branch of the Kern County Free Library, and the newspaper regularly printed the lists of new books that arrived at the local library. In the same year that *Polly of the Midway-Sunset* was published, other books reaching the Taft Branch Library included *Lord Jim* (Conrad), *Last of the Mohicans* (Cooper), *You Know Me, Al* (Lardner), *Little Lord Fauntleroy* (Burnett), *Uncle Remus and His Friends* (Harris), *Soldiers Three* (Kipling) and *Hans Brinker* (Dodge). Whatever the respective literary merits, *Polly of the Midway-Sunset* could claim one distinction. It was the first novel set in the Midway-Sunset oil field, and as such it touched a responsive chord of community pride.

In the novel, Petroleum City's school enjoyed such "an enviable reputation under management of the trustees and supervising prin-

Cox & Foster's soda fountain, circa 1912, offered a polite setting for dates in early-day Taft. (Photo from Clarence Williams Collection)

cipal . . . that educational speakers from near and far felt honored to be called upon to express before interested audiences their views regarding the ups and downs of the road to knowledge."

It was at one such meeting on a Saturday afternoon in spring that Robert Downing met Miss Polly Perley. It was a chance encounter occasioned by Downing's late arrival at the meeting and taking of a seat, without design, beside Miss Perley. The speaker that afternoon was the Riverton School superintendent, who in the course of his remarks stated, "Poor schools, in a large measure, are a result of the fact that many teachers are acting in that capacity while awaiting promotion to the marriage state."

Intrigued, Downing asked the person sitting on his right, who happened to be Miss Perley, "Do you believe that?"

"Yes, every word of it," she replied. Then, with an expression "as impenetrable as that of the sphinx," she gave her whole attention to the speaker.

Downing took the opportunity to look more closely. He liked

what he saw, and decided to pursue the subject further. That evening, he called on Miss Perley, finding her at home in the bungalow where she lived with her mother and father. While Mrs. Perley sat discreetly just beyond the curtained arch of the bungalow's living room, "absorbed in a magazine," Downing made small talk as he watched Polly do dainty needlework. Downing, an orphan since an early age, had been "reared by a bachelor uncle in a boarding house, and had never known the delights of a home." He watched fascinated as Polly's deft fingers fashioned a blue violet on a white ground. He cleverly turned the subject to what he wished to discuss, stating that teaching must be hard work and very monotonous.

Polly replied, "Hard work, yes. Monotonous, no. Almost every day brings us a bit of brightness."

Speaking of schools, Downing said, led him to think of books and literature. He asked who her favorite poet was.

When Polly replied Longfellow, he realized that she had played into his hands and took the opening to remark, "If you agree with Longfellow, you can't agree with the superintendent."

Polly reaffirmed her agreement with the superintendent, saying, "A successful working woman must center every thought on her work. She has no time for anything else."

Hardly ready to concede, Downing picked up a volume of Longfellow's poems that conveniently happened to be lying on a nearby table. He opened the volume to "Hiawatha" and read aloud:

"As unto the bow the cord is,
  So unto the man is woman;
  Though she bends him, she obeys him,
  Though she draws him, yet she follows,
  Useless each without the other."

Without a change in countenance, Polly coolly replied, "What you imply is not true for a working girl. Her head must not be turned by a lover, if she would succeed in her chosen profession."

"You don't honestly believe that?" Downing asked.

"I do," she rejoined.

Downing, unwilling to give up, asked, "Would you refuse to receive the advances of a young man, who might become interested in you, solely because his attentions would, rightly, require a part of your time?"

"Well," Polly replied, veiling her mischievous eyes from the young man, "the condition does look alluring, but I am still of the opinion that I would refuse his advances."

Disappointed, Downing soon afterward left. After he had gone, Polly Perley repaired to her writing desk and with some concentration composed a poem, which she put into an envelope and subsequently mailed to Robert Downing. The poem read:

"'As unto the bow the cord is,
  So unto the man is woman,'
Flowed these words in stately measure,
From the poet's gifted pen,
In the days when all creation
Bowed unto those tyrants—men.
And the subtle man of wisdom,
Knew if he would sell his verse,
He must sing of meek submission,
In a style both clear and terse.

"'Though she bends him, she obeys him,'
Where's the man, now tell me, pray,
Who would dare such thoughts to utter
To the woman of today?
He's not living on this planet,
He's now laid upon the shelf,
For the new progressive woman
Is a law unto herself.

"'Though she draws him, yet she follows,'
I maintain this is not so;
Walks she in the ways of reason,
Not the paths where man doth go.
For 'tis proven by statistics
That he often travels wrong,
And her faith is based on figures,
Not on sentimental song.

"'Useless each without the other,'
You may think these words are true;
But as a self-reliant woman
I take of them a different view.
I prefer to paddle single
Down the treacherous stream of life,
Rather than go with another
As a meek, obedient wife."

Pepper trees which bordered the fictional oil superintendent's house in *Polly of the Midway-Sunset* had their real life counterpart on many West Side leases, including General Petroleum's Nevada-Midway camp. (Photo from Clarence Williams Collection)

Left, a flaming well provided part of the plot for *Polly of the Midway-Sunset*. In real life, such events provided occasional excitement in the West Side oil fields. (Photo from West Kern Oil Musuem)

Sometimes the column of smoke and flames which motorists turned to watch came from a tank farm. Above, General Petroleum's stored oil goes up in flames. (Photo from West Kern Oil Museum)

If Miss Polly Perley had been given her choice of a time in which to declare her independence of a subservient role as a wife, she perhaps could not have chosen a better time than the year of 1917 when her fictional story appeared in print.

Concurrent with publication of the novel, the role of women was changing, not only in the oil fields but elsewhere as well. It was a point of some note in Taft that for the first time a criminal case had been tried in town with a "mixed" jury. The case involved the arrest of a man who ran the shoeshine stand next to the Shamrock pool hall on the charge of carrying a concealed weapon. The man had been arrested in the Boust City area with a revolver in his possession. He claimed he had taken the revolver from another man as security for a $5 loan and was merely transporting it to his quarters. In the docket, the defendant faced a jury that included eight men and four women. He told his story, and the jury acquitted him.

In Taft, women were making gains in the job field. It occasioned merriment in town when a war of words broke out in the Los Angeles press over the rival claims of the cities of San Diego and Riverside, each declaring it had been the first to hire women as Western Union messengers. The *Daily Midway Driller* smugly pointed out that the Western Union office in Taft had hired two female messengers a full month before either of the two larger cities. The messengers were Mrs. Katherine Dowd and Miss Florence Crawford. Things seemed to be working out well, with each sometimes getting as much as a $1 tip.

Nor were job opportunities confined to lower echelons. At the Taft Garage, Beth Berg, who had previously been the firm's secretary, became a new car salesman. She described herself as partial to the Overland Country Club roadster and advised oil men to give them to their wives for Christmas. After knowledgeably explaining the generator, lighting system, carburetor, cooling system, cantilever springs and excellent upholstery to Mark Broderick of the North Midway, she succeeded in making her first sale.

There was a certain amount of justice in the promotion of women to the field of sales, for they were as proud of their cars as men. Among those whose pictures appeared on the front page of the Taft newspaper behind the wheels of their new cars were Mrs. P. P. Means of Sanitex Cleaners in her "classy" Liberty Six, which sold for $1,195 and was described as "one of the neatest cars on the West Side, delivering comfort like the biggest car made," and Mrs. Ed

Craghill of the Alvord Hotel, who drove her "chummy Chalmers" as "a reminder that her hubby is a great sport."

Elsewhere, women were even taking an active role in the oil fields. The *Driller* reported that at Irvine, Kentucky, the 24-year-old Mrs. Daun Williams Spice was "the lady superintendent who wears the overalls." Mrs. Spice, the newspaper said, "backs up the smiles with action. She is on the job in all kinds of weather, seeing that drillers and tool dressers are at work on various leases now estimated to be worth more than $10 million." The newspaper said she was "not mannish nor a mean boss," adding, "The various rig crews, pumpers, gaugers and engineers obey willingly and gladly the lady with the blue overalls. The reader need not blush when it is said that the lady boss of an oil district wears 'breeches,' for convenience and comfort enter largely into the wearing apparel of all who are engaged in the business. Style is forgotten for comfort and it is nobody's business if Daun Williams Spice chooses to wear $1.50 overalls." The impact of the news story suffered somewhat by the further identification of Mrs. Spice as the daughter of M. E. Williams, who owned the field at Irvine.

Mrs. Spice was not the only woman in oil. The newspaper also reported the formation of Great Western Petroleum Corporation in Chicago by three young women stenographers. The company was

A carriage for two, Fourth Street at Center in Taft, circa 1912. (Photo from Clarence Williams Collection)

incorporated under the laws of Delaware, and it was said to have a capitalization of $5 million.

Though Miss Polly Perley was not doing anything so startling as organizing oil companies to keep her in the public eye, Robert Downing could not dismiss her from his mind.

His first reaction when he read the poem she sent was anger, but on reflection, soothed by "the soft breath of Spring laden with the wholesome tang of oil," he cooled down and blamed her attitude on "conditions of the times which allow girls so much independence of thought and action." For him, her attitude was one more proof of the folly of allowing women to vote.

It was inevitable that the fictional paths of Miss Perley and Robert Downing should cross again, even as they probably would have done if both had been real people living in the small town of Taft. The circumstances of their meeting involved the annual auto races in a nearby town identified only as the county seat. Even as Petroleum City sounded suspiciously like Taft, the county seat sounded even more like Bakersfield, which in real life lay some 40 miles away from the oil community and was linked to it by a highway that sounded like the one over which Downing and a friend rode to see the auto races.

Describing Downing's trip in the friend's automobile, the author wrote, "The ride was one of delight. A few miles out was Lake Buena Vista, shimmering and flashing in the sunlight, and curling its bright waters about the cool, green islands of tule which fringe its irregular shores." The trip took Downing and his friend past "the great tank farm, now dazzling to the eye in the radiant sunshine . . . reflecting into the blue of the air the myriads of whirling and dancing light waves." The beauty of the trip was enhanced "by the dark green of the alfalfa fields, bordered with the clear blue of irrigating canals, over which arching sprays of glossy foliage served as perches for the singing birds, when wearied of dipping and darting deliriously in the intoxicating air."

The euphoric mood was jolted after Robert Downing and his friend took seats in the grandstand. In the crowd, Downing saw Miss Polly Perley, accompanied by her mother and a man Downing recognized as a successful oil superintendent, a "middle-aged, good-looking" man whom Downing knew to be a bachelor. Was the superintendent interested in Miss Perley? The thought was dismaying. The man had "good looks, brains, education, culture and a

Community pride in the West Side oil fields evidenced itself in parades like the Fourth of July parade in Taft in 1912. (Photo from Clarence Williams Collection)

princely salary." How could Robert Downing compete?

Downing was jarred from his somber musings by the roar of the crowd watching the great Barney Oldfield circle the track. In real life, Oldfield was a frequent competitor in Bakersfield races. Putting thoughts of Miss Perley from his mind, Downing settled back to watch the races. After they had ended, he ate dinner with his friend at a local hotel after which they drove down Chester Avenue. At the "unique stone clock tower," Downing "bade his companion halt for a moment to view this attractive memorial, hinting in its style and setting of old world art and custom." After the clock tower had been duly appreciated, the two men set out for the return drive to Petroleum City.

Enroute, the peaceful evening was shattered by the sight of a burning oil gusher, lighting the sky. "Up and up the scarlet column mounted, spreading umbrella-like as it rose, and showing patches of silvery clouds, signals of the heroic workmen vainly striving to smother the fire god." Even as Downing confidently predicted to his friend that the well would be brought under control, "knowing the wonderful energy and determination of the workers of the oil fields," the friend devoted more attention to the spectacle than to his driving, and their trip came to an abrupt halt when they plowed into the rear of another vehicle. Fortunately, there were no injuries, but the accident did lock together the two cars, forcing the occupants to dismount and wait for help. Embarrassingly, the other car was one driven by the successful oil company superintendent and his passengers were Miss Polly Perley and her mother.

After a brief exchange between the two drivers that threatened to become recriminatory with charges that each had been watching the burning well with undue attention, and which was quickly subdued by the presence of the ladies in the superintendent's car, the occupants of the cars passed the time in amicable conversation.

In the course of the discussion, Miss Perley pointed out that things could have been worse. By way of elaboration, she told them of a little girl in Petroleum City, "a dear, tiny girl of six who, through illness, which has seriously affected her baby eyes of blue, is threatened with total blindness." Miss Perley's voice broke as she reflected on the girl groping in vain for light.

Fortunately, there was a ray of hope. "A certain delicate operation may save her sight," Miss Perley related. It seemed there was "a wonderful surgeon" in New York who had agreed to come to the oil field for $2,000 to perform the operation. Unfortunately, the child's loving parents were so poor they could not afford the fee. They were so destitute that the mother had to work in a laundry.

The oil company superintendent had a ready response to the problem of raising the money necessary for the operation. "When did the men of the oil fields ever turn a deaf ear to the cry of distress?" he asked.

Robert Downing was quick to respond, too. "We who live in Petroleum City want also a share in helping to make up that purse."

Later, Downing reflected on Miss Perley's concern for the little girl and decided that women should not be barred from voting after all. Because of women like Miss Polly, he decided, the world was "a better and more attractive place." He consoled himself that the superintendent, too, had conducted himself "right manly." Miss Perley's future was safe with him, he sadly concluded.

If some of Miss Michaels' readers outside the oil fields might have wondered at the superintendent's confidence in quickly raising the necessary funds, none in the Midway-Sunset field would have regarded his response as anything but matter-of-fact.

When the novel appeared in 1917, the nation was engaged in the war in Europe, and each day seemed to bring a new campaign in Taft to raise money or other aid for those who needed it. Through the efforts of Mesdames L. P. Guiberson, H. A. Hopkins and H. H. McClintock, wives of prominent local oil men, the people of the

Young men and ladies of Taft enjoyed an outing on horseback into the nearby Temblor Hills and Elkhorn Valley in March 1911, leaving Lierly & Son's North Midway Stables at 8:30 A.M., returning at 6:00 P.M.. (Photo from Clarence Williams Collection)

West Side had contributed generously to Belgian relief, shipping several hundred pounds of clothing, shoes, hosiery and other garments to Los Angeles to be forwarded to the stricken Belgians. Mrs. R. W. Patterson was heading a campaign to raise funds to supply libraries for soldiers in the United States. The local chapter of the Women's Christian Temperance Union was leading a drive to raise $1,650 to purchase and equip an ambulance for the Red Cross for service in France. Teachers in the Browngold School at Shale, seven miles from Taft, did their bit by raising flowers in the spring, mostly sweetpeas, which they sold in the neighboring oil fields, earning $25 for the Red Cross. Those raising and selling flowers included the school's principal, Mrs. Curl, and its two teachers, Miss DeVana and Miss Fletcher.

Almost every week seemed to bring a benefit performance of one sort or another to two local theaters. For the benefit of the Red Cross, the C & C Airdrome featured a benefit showing of the movie *Womanhood*, starring Alice Joyce. The movie, advertised as "another Joan of Arc picture," promised viewers such scenes as these: "A pacifist meeting turns into a riot that is anything but pacific; a heroine of the nation is kidnapped in an aeroplane by foreign agents;

a whole navy is destroyed in a sea of burning oil; gas attacks as conducted on the battlefield are reproduced with fidelity to actual conditions." The Optic Theater offered the Berne Troupe in "a marvelous demonstration of the tricks of self defense as practiced by the people of Lapland since the Eleventh Century." Admission was 35 cents for adults, 15 cents for children.

No campaign met with quite the enthusiastic response of the drive that went by the name "Smokes for Sammies." The campaign involved the solicitation of money to buy smoking kits to be sent by the American Tobacco Company to American soldiers on the western front. Each 25 cent contribution paid for a kit that included two 20-cigarette packs of Lucky Strikes, three sacks of Bull Durham smoking tobacco, one can of Tuxedo smoking tobacco, six books of Bull Durham cigarette papers, and a postcard for acknowledgement of the gift by the recipient. The kits represented a bargain, with tobacco inside normally retailing for 45 cents.

The *Driller* headlined the campaign with the slogan, "Load up the pipes of the boys in France." With lists of contributors, the newspaper on its front page ran stories detailing the liking of the boys at the front for smokes. "Of soldiers waiting to go into the firing lines to fight for justice and liberty," the newspaper reported, "it can safely be said 99 percent smoke."

It was one thing, the paper reported, to light up a cigar in the comfort of one's home. It was quite another "to light a cigarette when you're sitting on the fire step in a front line trench wondering about your next expedition 'over the top' into No Man's Land." Noting the difficulty of securing American tobacco overseas, the newspaper said, "That curly hot French tobacco is all right for the *poilus*, but not for our men." The newspaper reported that the War Department had lent its full endorsement to the campaign, told of army officials' agreement that smokes were necessary for fighting men, quoted General Leonard Wood at Fort Riley, Kansas, saying, "Nothing gives a soldier in the field more pleasure and contentment than a cool refreshing smoke after a hard day's fighting," and also quoted an unnamed private as saying he would "a darn sight rather have cigarettes than socks." The newspaper added, "Men of science have known for years cigarettes are as harmless in moderation as many other things that we all do regularly." The paper pointed out that A. G. Empey, author of the best-seller, *Over The Top*, wrote of a stretcher-bearer whose dialogue with wounded men normally

Mercantile store offered everything from cut glass and rugs to hammocks and fine cutlery in the up-and-coming city of Taft, which was the model for Petroleum City in the novel *Polly of the Midway-Sunset*. (Photos from Clarence Williams Collection)

went, "Want a fag? Where are you hit?" Less than two months after the "Smokes for Sammies" campaign had begun, West Side people had sent more than 2,500 tobacco kits to soldiers in France, a number greater than the population of Taft.

Even as oil people raised money for the various causes of 1917, the fictional characters of *Polly of the Midway-Sunset* succeeded in raising the money that was necessary for the operation to save the little girl's sight. The fund-raising effort was to have a passing effect on the love affair between Robert Downing and Polly Perley.

On a particular day, word came to Downing that the successful oil company superintendent had just returned to Petroleum City after having been married, a marriage about which Downing had not previously heard. Disturbed by the thought that Miss Polly was no longer among the ranks of available girls, Downing went for a walk that evening in the oil fields. While walking, he was overtaken by the superintendent in his car. Funds having been raised for the operation, the superintendent asked that Downing accompany him to his house to get the money, which the superintendent, because of the press of business, felt he would not be able to take to the bank the following day.

Reluctantly, Downing accompanied the superintendent after the latter assured him his wife would be happy to receive him in their house, even though she had no warning a guest was coming. At the superintendent's lease house, Downing "noted with approval the border of pepper trees with their delicate frond-like foliage, and the trailing vines covering the screened veranda which, with its roomy coolness, enclosed the house." The superintendent's wife was not at home, and he insisted Downing wait in the living room, "where books bespoke a man of culture," while he went to seek her. Shortly, the superintendent returned with Miss Polly Perley and another woman, Polly's aunt. Much to Downing's surprise, it developed that it was the aunt, not Miss Perley, who was the superintendent's bride.

Love would be denied no longer. Courtship followed, and with it an assist from the superintendent, who on a June day took Robert Downing with him and his wife to visit at the ranch of his wife's sister and her husband. Among those present were Miss Polly Perley, the surgeon who had successfully performed the operation to restore the little girl's sight, and other relatives. In the afternoon the group gathered on the porch, "filled with the blooms of the flowering gardens, and rustic dishes heaped with apricots, cherries and early peaches." A neighbor jolted the pleasant conversation with the remark, "Why is it that one sees so often nowadays the Modern Girl trotting in single harness?"

Miss Polly replied, "It may be that the Modern Girl is tired of listening to, 'I wish, my dear, you could learn to make such pies as my mother used to make.' Possibly she is going to postpone matrimony until all food is chemically prepared."

The surgeon could not resist joining in, stating, "Nothing will keep the savage ancestral blood of the future husband fron surging to the surface, and he will cruelly assert, 'Man is the superior of woman.'"

To which Polly replied, "Then the future wife will serenely say, 'Then the ordinary clay of the street is finer material than that into which it was transformed in the framework of man's wonderful body.'"

The conversation ended when the time arrived for the surgeon to catch the train for his return to New York. In parting, he held Polly's hand in his—longer, Downing thought, than etiquette demanded—and said, "Homekeeping hearts are happiest." Love-smitten, Downing thought, "Fortunate the man who wins Polly."

The moment for his proposal came later, in a canoe on the waters of Lily Lake, "a beautiful sheet of water set in shores that gradually rise in hills of living green." There, "in entire forgetfulness of the dreaded 'no' that might come, irresistibly impelled by a strong wave of feeling," he said earnestly:

"There is a woman's heart that if I might win 'to have and to hold,' I would cherish it most sacredly."

As Miss Polly's heart throbbed with sudden emotion, Downing took her into his arms and "ventured a light kiss on the lips so temptingly near his own."

Even as she gave him her happy answer, he had one more question. How had she happened to write the words:

"I prefer to paddle single
Down the treacherous stream of life,
Rather than go with another
As a meek, obedient wife."

With sudden recollection, Polly said, "I wrote the first two lines simply because they came into my mind. Than I wanted a word to rhyme with 'life,' and 'wife' was the first word I thought of. Wife does rhyme with life, doesn't it?"

"It is the only word that does," exultingly exclaimed the happy Downing.

Discovery on the School Section

The section of land was one the federal government gave to the state of California in 1853 three years after the state had been admitted to the union. The grant included the sixteenth and thirty-sixth sections of each township. Its purpose was aiding development of the state's school system. There was a qualification to the gift: the federal government retained all "known mineral land." The state largely had had its inception in the great rush that followed the discovery of gold five years before by James Marshall in the tailrace of John Sutter's sawmill on the South Fork of the American River. The Congress of the United States wanted to support California's schools, but not to the extent of giving away a Mother Lode.

The particular section seemed at the time it passed into the hands of the state to be too poor a candidate to ever make any substantial contribution to schools, or for that matter to anyone or anything else. It was situated in a low range of hills in the southern end of the San Joaquin Valley, many miles removed from Mother Lode country or known occurrence of any mineral that made a significant contribution to the well-being of the country. Any thought of agriculture was out. The annual rainfall was seldom more than a few inches, scarcely enough to support crops. The section was in hills where irrigation would be impossible. For the time being, the section would not be more than a treeless patch of ground where sagebrush grew and weeds turned brown in summer sun. The most life it would see would be the native elk that wandered up from the tule-infested shores of nearby Buena Vista Lake in such numbers that they eventually gave their name to the hills in which the section of land was located.

Fifty years after the federal government gave the section to the state, the state realized its first return. It was not a magnificent sum that would make much difference in the financing of the state's school system. In 1903, the state sold the land to Alice Miller for $1.25 an acre, giving the state a return of $800. A short time later Alice Miller let the section revert to the state in lieu of paying taxes assessed on it. The tax bill was $3.82.

A few years later in 1908 the state found another buyer, selling the section for the same price of $1.25 an acre, realizing another $800 to be spent for school purposes. The new buyer was a man named Hay who had happened along at a time when oil prospectors were busy in the southern end of the San Joaquin Valley. A year after he had purchased the section, he resold the property, getting $20 an acre, which enabled him to turn a handsome profit. The purchaser was Standard Oil Company.

Standard, which had entered the California oil picture early, was interested in expanding its holdings. The company bought the property for speculation, regarding the purchase price of $20 an acre, a total of $12,800, as an acceptable amount to pay for a land position in an area which, though not yet proved productive, was only a few miles from production underway in the Midway district. Having acquired the property, Standard did nothing. The company sat tight, content to wait until there seemed some compelling reason to do something with the land. The section was Sec. 36, 30S-23E, Kern County. It was in Elk Hills, and because of the circumstances under which the federal government had given it to the state, it came to be known as the "school section."

The same speculation that brought Standard Oil to the school section also attracted another party to Elk Hills, though not for precisely the same purpose. The party was the United States Navy. The Navy's interest stemmed from the pioneering work of men like William Matson of Matson Navigation Company in San Francisco. As early as 1901 Captain Matson had converted the *Enterprise* from a coal-burning to an oil-burning vessel and steamed her to the Hawaiian Islands, proving that oil could power a ship at sea as well as, if not better than, coal. Coal-burners had always been plagued by the spectre of running out of fuel during adverse seas. Because of the much greater amount of energy possible from equivalent amounts of fuel oil, the switch to oil promised to eliminate the danger. The Navy early recognized the advantages of the new fuel and looked around to see how available it might be.

The time was the early 1900s. It was a time when public lands throughout the west were fast being transferred to private owner-ship, mainly through claims which had ripened into full private ownership. Even as the Navy made plans to make a complete changeover from coal to petroleum, and also for an expensive new ship construction program based on the use of oil, Navy planners

Left to right, Ford Alexander, Clarence Hendershott, Alexander's brother-in-law, and K. T. Kinley with five-foot-long torpedoes, each filled with 100 pounds of dynamite, used to snuff out well fires. Kinley's sixteen-year-old son, Myron, not shown, who sometimes worked with Alexander and the others, later became the oil fields' premier tamer of wild wells. (Photo from Boyd Alexander)

realized that they might have to pay a high price for oil to power their ships because the government had lost control of all the potential oil lands.

The suggestion forthwith was made that probable oil-bearing land in the public domain should be permanently withdrawn from areas on which claims could be staked. President Theodore Roosevelt, who believed in talking softly but carrying a big stick, was quick to realize that the big stick might not be effective if it did not have fuel to keep it afloat. He subsequently directed the United States Geological Survey to inquire into and report on those parts of public lands believed to contain oil.

George Otis Smith, director of the U.S.G.S., shared the President's views. Under his enthusiastic direction, the survey was quickly begun. Smith, in a letter of February 24, 1908 to the Secretary of the Interior, called the Secretary's attention to the superiority of petroleum products to power steamships, and also advised of the British government's use of such fuels as emergency fuels in battleships. On that account, Smith wrote, it was his recommendation that filing of claims to oil lands in California be suspended. "If we don't stop it,"

Smith wrote, "the government will be required to repurchase oil it has practically given away." Smith pointed out that only a limited area with oil potential was left in California, and it was rapidly being filed on and patented, either through legitimate oil development or by subterfuge, over claims for gypsum and other minerals.

The U.S.G.S. survey was not completed until after President Roosevelt left office, but early in the administration of President William Howard Taft, the U.S.G.S. recommendations were made. On September 27,1909, President Taft signed an order which temporarily withdrew large areas in California and Wyoming from entry and settlement under existing public land laws.

The withdrawal order was promptly attacked. It was argued that the executive arm of the government could not constitutionally suspend laws enacted by Congress permitting acquisition of lands in the public domain, and that therefore the order was void. President Taft responded by asking Congress to pass enabling legislation, and Congress dutifully passed the Act of June 25, 1910, known as the Pickett Act, which vested the President with discretionary power to make temporary withdrawals. The statute expressly preserved the rights of any person who on the date of the withdrawal order was a bona fide claimant or occupant of oil and gas lands. President Taft by executive order of July 2, 1910 confirmed the withdrawals he had made earlier. In neither of the withdrawal orders was the Navy mentioned. Lands were merely withdrawn from private entry and were left under the jurisdiction of the Interior Department.

On June 25, 1912, the Secretary of the Navy asked the Secretary of the Interior for cooperation in securing the reservation for the Navy of oil-bearing public lands in California sufficient to insure the supply of 500 million barrels of oil. In response the U.S.G.S. recommended an area of 38,072.71 acres at Elk Hills. Accordingly, President Taft, on September 2, 1912, issued an executive order creating Naval Petroleum Reserve No. 1. Of the area lying within the boundaries of the reserve, 12,103.09 acres appeared to be legally owned by private owners, including the school section which belonged to Standard Oil Company. The balance of 25,969.62 acres remained in government ownership.

At the time, no actual discoveries of oil by drilling had been made. Selection of the area was founded mainly on general knowledge of geology. No one knew whether the area contained more or less than the 500 million barrels of oil the Navy had requested, or if

Ford Alexander, in asbestos suit, tackled the wells that got away. Alexander quoted a price of $1,000 for extinguishing oil and gas fires, adding that no charge would be made if the fire was not extinguished. (Photo from Boyd Alexander)

it contained any oil at all. It remained to be seen whether the U.S.G.S., acting for the Navy, or Standard Oil, for that matter, had made a good choice.

It was not until 1918 that Standard Oil Company turned to the school section to see what, if anything, was there. The circumstance that led the company to the section nine years after acquiring it was the common one by which the oil industry historically had been motivated. A shortage loomed. If no other reason had existed, the World War, recently concluded, had convincingly demonstrated the importance of oil. As plain as it was that oil would become even more important, it was just as clear that oil was not as abundant as might be desired.

Standard, for one, recognized the problem, and took out half-page advertisements in various newspapers to point out the urgent need for action. The headline over the advertisement read, "The Gasoline Problem of Supply and Demand." In the text below, the ad stated, "The disproportion between the supply of oil and demand for gasoline is enormous and constitutes a critical problem.

"Projected at the percentage of increase, 1904-1914, we should require in 1927 something like 700,000,000 barrels of petroleum. In 1918 our total production was only 350,000,000 barrels."

Having thus summarized the dimensions of the problem, the ad went on to state that both petroleum and automobile industries had for several years been making every effort to meet the situation. On the part of the oil industry, the effort included "constantly prospecting for new fields." For the automobile industry, it included setting standards for gasoline. In regard to the latter, Standard indicated its willingness to cooperate in the worthwhile effort by concluding its advertisement with the statement, "All Red Crown gasoline now being supplied in the Pacific Coast states is refined to conform with the United States Government standard specifications. It has the full, uniform chain of boiling points necessary for full-powered, dependable gasoline: low boiling points for easy starting, medium boiling points for quick, smooth acceleration, and high boiling points for power and mileage."

As for prospecting for new fields, Standard was doing its part. The company in 1918 was in the process of drilling what would amount to a total of 98 new wells in the state of California. The search was going on against a background gloomily summarized by a U.S.G.S. report that said the United States must bring in new oil fields very soon for it was believed that over half of the oil in the ground had already been consumed. It was not surprising that the search for new oil led Standard Oil Company to Elk Hills and to the school section.

Not all were agreed that Elk Hills was a good prospect. Associated Oil Company, for one, did not think much of the area. An Associated geologist had visited what he described as an "alleged" Elk Hills oil seep in 1909. He had described it as "a small amount of black tarry substance in patches partially exposed" and concluded that it was not an oil seepage "but an occurrence of organic material, probably plant remains." He had said, "I concluded that the Elk Hills might have small scattered deposits of oil, but that they would not be important in an economic sense." There had been some shallow wildcatting in various parts of the Elk Hills, always with the same dry results. An estimated three to four million dollars had been spent by various companies in fruitless drilling.

Standard was confident. Of oil companies doing business in California, Associated had more wells, 1,048 compared to Standard's 771, and the Southern Pacific Company's Fuel Oil Department had more proved land, 18,267 acres compared to Standard's 8,187 acres, but Standard had the edge where it counted. The company produced 22.6 per cent of the state's total production of 100 million barrels a year, compared with Associated's 9.1 per cent and Southern Pacific's 8.5 per cent.

Flames from runaway Hay No. 7 turned night into day at Elk Hills, August 1919. (Photo at left from Phil Witte; at right from E. W. Brubaker)

In the fall of 1918, Standard Oil Company built a wooden derrick and moved rotary drilling tools to Elk Hills to drill a wildcat on the school section, nine miles north of Taft. The project was under the supervision of Lin Little, who was known as "Thrift" Little in Taft because of his success in selling Liberty Bonds during the drives held in the West Side oil fields during the war. He was at the time finishing up a sale in which Taft citizens had purchased $236,098 worth of the bonds, substantially oversubscribing their quota of $200,000.

The first well was designated as Hay No. 1. Drilling progress was steady. In due course, the *Daily Midway Driller* in its edition of January 11, 1919 reported that Standard Oil had struck oil at Elk Hills. The newspaper said the Hay No. 1 flowed 200 barrels a day of high gravity oil from 2,480 feet. The production was hardly spectacular, but it was of interest for at least one reason. "This well has been a center of interest since work was begun on it," the newspaper reported, "as its outcome would prove whether the contentions of Secretary Daniels (of the Navy) as to the potentiality of the famous Naval Reserve were justified or not." The newspaper concluded, "The territory is proved valuable."

The find at Elk Hills was overshadowed by a bulletin from the Nebraska statehouse that the state had become the 36th to ratify the prohibition amendment. The amendment decreed the end of alcohol in the United States, and its passage led prohibition leaders to describe the accomplishment as "the greatest piece of moral legislation in the history of the world." At the same time the United States was going dry, Elk Hills had its first well.

If reaction to the discovery at Elk Hills was less than exuberant, it could easily be understood for the West Side was used to bigger things than 200 barrels-per-day wells in its oil fields. The discovery well hardly compared with some being brought in a few miles east of the town of Taft. Honolulu's No. 19 on Sec. 10, 32S-24E at Buena Vista Hills roared in a few days after Standard completed the Elk Hills discovery well, throwing oil over the derrick with a flow estimated at 7,000 barrels a day. The Honolulu well underscored the West Side's dominant role in the California oil picture. Soon afterward when production figures were released for the recently ended year of 1918, Midway-Sunset led the state, having produced 34 million barrels during the year. The closest competitor was Brea-Montebello-Whittier in Southern California, which was credited with 25½ million barrels.

Robert H. Nicol, valley editor of the *Fresno Republican*, visited Taft and wrote on his return to Fresno, "A visit paid during the current week to Taft, the capital of Midway, revealed a city of 3,000 inhabitants, nestling close against the sand hills of the desert, and while it modestly claims but 3,000 of population, tributary to it from the camp leases surrounding it are fully three times that number. One is astonished after a ride of thirty miles over a finely paved highway from Bakersfield, and over vast stretches of desert, to find a modern city, and not alone that, but an immense forest of derricks, every one of which represents the outlay of a fortune expended and later doubled and sometimes trebled for the promoters.

"Taft has electricity, water, gas, fine fire department, a clubhouse, all lodges represented, good hotels and everything for the comfort of the dweller and visitor. It has a moving picture theater that cost $30,000 to equip owned by C. L. Langley and F. W. Livingston, which has the costliest equipment of any theater on the coast, it is claimed, and will seat 1,500 people. Stage lines leave the city every hour for the big cities, and a railway brings a train in once a day and departs similarly.

"The number of producing wells in the Midway field at present is 2,120 with nearly 100 new ones drilling. The daily labor output on each drilling well is $105. The average cost of each well is $30,000.

The world's most productive gas well, Hay No. 7 at Elk Hills, after the blazing well was brought under control. (Photo from John C. Maher, United States Geological Survey)

"There are six pipe lines laid to the coast, for carrying oil. The natural gas production is unlimited, and now supplies Los Angeles. The total production of the field to date is 900,000,000 barrels of oil.

"Workmen receive good wages, from $5 to $15 per day, and most of them are skilled workmen...The year 1919 is destined to be a record one for the Midway."

Another visitor was a man named Fred H. Hamilton who could hardly wait to get home to Rensselaer, Indiana, to give the *Rensselaer Republican*, another newspaper, his impressions of Taft. He forwarded the write-up to friends in the oil fields. Hamilton was quoted as saying, "Taft, California, is situated in the heart of the richest oil well country in the world. It is surrounded by small mountains on each side and any time of the day the snow-capped peaks can be seen. The Standard Oil Company has the largest holdings out here and, of course, there are smaller concerns.

"The city of Taft has a floating population and at the present time it is about six thousand. The homes here are called shacks. They are all in one story and have from three to eight rooms in them. Gas is very cheap and due to the wonderful weather they have here the year around they do not have to invest in coal.

"There are more automobiles and money here than any other place of its size. The wage scale is very high. The lowest paid is $4.00 per day and from that up, well, the sky is the limit.

"The city puts one in mind of a western movie city. The only thing it lacks is the bar and of course the cowboys riding on their wild horses. They make up for the wildness by the speed they travel in their cars.

"They have wonderful highways here and, of course, there is a speed limit but it is broken every day, every hour and every second...It is a great sight to see a $2,000 automobile parked in front of a home that does not cost over $300."

If there was no observed speed limit for cars, there was also no lack of speed with which Standard moved ahead to develop the school section. Normally when a wildcatter discovered an oil field, the hard days were over. This was not to prove the case for Standard. Just when it appeared the company could sit back and happily develop an oil field, bureaucratic trouble intervened. It came not from the Navy, Standard's neighbor at Elk Hills, but from another source, a relatively new agency of the state government. The agency was the Department of Petroleum & Gas, which was the forerunner of the Division of Oil & Gas.

In California, oil operators had come first, drilling with no particular interference from government agencies, regulated only by their ability to get leases or buy land, to secure rigs and men to run them, and to produce oil and sell it. The period of every man for himself continued through the end of the nineteenth century and into the early years of the twentieth, coinciding with the earliest and most unsophisticated days of oil activity, when men drilled on seeps and there was little understanding either of the forces that trapped oil or of the technology of producing it. Like other expanding industries, the oil industry became more knowledgeable as it went along, and the running of things little by little began to pass from the hands of men with derring-do into the hands of men who had gone to college and studied engineering, geology or other sciences. As the body of knowledge grew, it became apparent that oil wells were being ruined by poor production practices.

Through the years the State Mining Bureau had kept a relaxed eye on what was happening. Its duties consisted principally of gathering and publishing information about the various mineral resources of the state rather than attempting to regulate the oil industry. In 1914 the Bureau decided to make the first comprehensive survey of the California oil industry. The man picked to do the job was Roy McLaughlin, a mining engineer whose first introduction to oil had come unknowingly when as a youth in a Colorado mining camp he had stood by the back of a traveling man's wagon and listened to the huckster extol the marvelous qualities of a potion he claimed would relieve many human ills and actually cure deafness. The huckster called the substance Romany oil, and he sold it in pint bottles at a price of one dollar each, or three for two dollars. The Romany oil had a distinctive odor that McLaughlin would recognize later when he saw his first single cylinder gasoline engine and smelled the odor of the fuel on which it ran, identifying gasoline and Romany oil as one and the same. McLaughlin had worked in mining camps like Bodie, California and Manhattan, Nevada. When the mining boom burst, he had gone to work as a geologist in oil fields at Taft. The survey that he made of California's oil industry for the state revealed that in many fields damage to oil-bearing strata was caused by infiltrating water that had not been properly shut off.

An outgrowth of the finding was the formulation of legislation that was signed into law by Governor Hiram Johnson on August 9, 1915. The law, declaring that the people of California had a proprietary interest in the deposits of oil and gas under the state, set up regulations to protect such deposits against physical waste and established the Department of Petroleum & Gas in the State Mining Bureau. Governor Johnson picked McLaughlin to be the first State Oil & Gas Supervisor. McLaughlin selected as his right-hand man in the southern San Joaquin Valley a geologist named Roy Ferguson.

The Standard operations at Elk Hills fell within Ferguson's district. After a detailed study, Ferguson became convinced of the existence of a gas zone in the locality and recommended that casing be landed at a considerable depth above where Standard expected to encounter oil. Standard paid no heed to the recommendation, and managed to drill the second and third wells through the gas without recognizing its existence, getting a 400-barrels-a-day well at the second well, and an 850-barrels-a-day producer at the third.

An increasingly nervous Ferguson watched the development. When Standard proposed another well, he enlisted McLaughlin's support. Ordinarily McLaughlin left final decisions to the field deputy, but in this case he joined Ferguson for a meeting with Standard officials and the two state men insisted that the next well

Hay No. 8 at Elk Hills blew out and caught fire, too. (Photo from Kern County Museum)

should land casing to protect the gas zone. The state law lacked teeth to enforce the recommendation. After an angry discussion, McLaughlin and Ferguson agreed to make no further protest to Standard's plans for the next well so long as the well after that would be drilled in accordance with their recommendations. The compromise was accepted by Standard's manager of production, who assured McLaughlin and Ferguson in no uncertain terms that Standard Oil Company knew how to drill wells and that after completion of the agreed-on wells, the company would expect no further suggestions from them.

Early one morning in June thereafter McLaughlin was on the night train bound for Bakersfield where he was to inspect a well that was approaching critical depth. He raised the shade in his Pullman berth and saw on the distant horizon a plume of flames and smoke rising from Elk Hills. He knew without being told that nature had seconded the caution he and Roy Ferguson had urged on Standard. The Standard well had gotten away from around 2,000 feet and was flowing an estimated 30 million cubic feet per day of gas. It took some four days for the company to bring up the dozen boilers necessary to generate the columns of steam required to smother the fire. Afterward, the well was capped, and when lines were ready, put on production supplying gas for the city of Los Angeles.

Thanks to the blazing gasser, it was no longer necessary to try to convince anyone that Elk Hills was a significant discovery.

Land values skyrocketed. H. I. Tupman reported that he had been offered $1,500 an acre for 60 acres of patented land on Sec. 16, 30S-23E, four miles northwest of the Standard wells. He had purchased the land for $50 an acre. He did not identify the company making the offer, but General Petroleum Company soon afterward started a well on the section. The well ultimately proved dry at 3,474 feet.

Standard announced an aggressive exploratory program. The first follow-up wildcat would be on the Packard property, four miles southeast of the discovery. The drill site was on Sec. 16, 31S-24E, at the base of Elk Hills immediately adjoining the one and one-half million-barrel tank farm that Standard had completed three years before. There had been some unsuccessful drilling in the area in 1910, but the earlier wells had stopped at about 2,000 feet. Standard said if necessary it would go as deep as 4,000 feet at the new wildcat. Ultimately the well would prove dry at 6,240 feet. Nor was the company forgetting a development program on the school section. Among the new wells started was one designated as Hay No. 7. It was in a draw less than one-quarter mile from the original discovery well, which continued to flow at a good rate.

In the columns of Taft and Bakersfield newspapers, accounts of Elk Hills developments shared space with stories of Standard's ambitious plans to drill a wildcat projected to cost $100,000 on a 5,200-acre lease block near Moclip, Washington; the formation of the Imperial Valley Oil & Development Association by 200 bankers and businessmen intent on finding oil in Imperial Valley; and the acquisition by Petroleum Midway Company of an oil and gas lease on 5,000 acres on the California coast owned by Ignace Paderewski, the famed pianist and former premier of Poland who had originally purchased the property for a Polish colony.

Not all of the news was good. A Johns Hopkins doctor reported that the tobacco habit was dangerous. The West Side Drug Store advertised Nicotol tablets to help people kick the habit. From Pismo Beach came word that oil bilge was killing the famed Pismo clams. And from Coalinga came ominous news of the appearance of radical posters in the oil fields urging men to "Be a Red." The posters were thought to be the work of the Coalinga I.W.W., and a deputy marshal was quoted as saying, "The average person in America, having no cause for discontent, does not realize that a certain vicious element is trying to bring on universal strikes and a revolution." A half-page advertisement in the *Daily Midway Driller* offered a remedy for the threat of Bolshevism. "It has been said," the ad stated, "that those who term themselves Bolshevists and red flag artists are that way because they have no home." The solution, the ad suggested, was home ownership. "Build homes, remodel old ones, paint, repair and beautify," the ad said. Sponsors of the advertisement, self-described as "various public spirited citizens co-operating and lending their support in making this educational Home Building page possible," included firms and individuals offering services and materials necessary for home-building and beautifying.

In Taft, plans were progressing for a three-day celebration of the Fourth of July. It was to be a "welcome home" celebration for the men home from France. It would include auto races and motorcycle races, greased pigs and greased poles, boxing bouts and a "Days of '49" camp. Troops were landing almost daily in New York, and the men of the 91st Division's 363rd Infantry Regiment, largely recruited in the west, had only recently paraded down Market Street in San Francisco to the shouts of "Argonne, Argonne, Powder River, Let 'er Buck."

A coming attraction at the Hippodrome Theater was Harry Ray's eight professional shimmie dancers who would demonstrate the dance craze that was sweeping the nation. "Oh, Minnie! Shimmie for Me!" the advertisement read.

On the school section, Standard's crews continued to drill ahead.

Around midnight on Saturday, July 26, the crew on Standard's Hay No. 7 had reached a depth of 2,140 feet when the well blew in with a thunderous roar that could be heard clearly in Taft, nine miles away. Gas blasted into the sky, accompanied by a shower of rocks and dirt. There was nothing the crew could do but get out in a hurry. Fortunately no one was injured. Within an hour the friction of rocks striking against casing had caused a spark. Gas caught fire, burning fiercely in a torch that could be seen from fifty miles away.

It was an awesome sight. Men who only a few weeks before had battled the flames at another Hay gasser began preparations to contain the new gasser. Flames shot three hundred feet into the air,

Elk Hills in the 1920s. Compressor Station B, Midway Gas Company, circa 1927. (Photo from John C. Maher, United States Geological Survey)

and it appeared they were being fed by substantially more gas than at the earlier well. Standard rounded up all available men and began moving in boilers to mount an attack. It was estimated that gas was escaping at a rate of fifty million cubic feet daily.

The towering flames in the night sky attracted attention throughout the fields and in the neighboring towns of Taft and Bakersfield. On the following day, which was Sunday, motorists flocked to see the spectacle, clogging bumpy oil field roads and making more difficult the task of moving in and rigging boilers. Standard quickly built up its task force. Less than twenty-four hours after the well had gotten away, some 125 men had been assigned to crews laboring to control the wild well. There were three shifts of eight hours each. Work went on around the clock. All drilling operations were shut down on the Hay property.

The *Bakersfield Morning Echo* said of the spectacle, "It is the largest flame ever seen on the West Side. The roar of flames can be heard in Taft, nine miles away. The western sky at night from Bakersfield is alight with the mass of fire that spreads over the horizon. The volcano shoots about three hundred feet in the air. Flames are about three feet wide at the base but explode in fan shape as they shoot into the air so the great pillar of fire is probably twenty-five feet wide at its greatest diameter. The heat is intense and one can not come closer than three hundred yards."

Early development of Elk Hills field, 1920s. (Photo from E. W. Brubaker)

As if normal July heat of the southern San Joaquin Valley were not enough, to it was added the intense heat of the burning gas, creating an inferno in which to work. Workmen approached the flames like knights stalking a dragon, moving into the inferno behind corrugated iron shields twelve feet in height and twenty-five feet wide. The uniform was woolens. The combination of July heat, burning gas and woolens would, of course, have been intolerable if the team effort had not also included men farther back who manned hoses pouring a constant stream of water on the men who labored to set up boilers and rig lines so that steam might be turned against the well.

By Tuesday, the third day after the blowout, ten boilers ringed the gasser. Standard officials decided it was time to go on the attack. Though the flames had been roaring into the sky for some three days, they showed no sign of any loss of pressure. Top joints of casing had been blown from the ground, and the fire appeared to be burning from about twelve feet down. When all was in readiness, the signal was given. Ten jets of steam were played on the fire. Of the effort, the *Daily Midway Driller* reported, "A battery of ten boilers failed to make any impression on the huge column of burning gas." A call went out for more boilers.

By Friday, six days after the uncontrolled blow had begun, sixteen boilers ringed the gasser. Far from weakening, the flames if anything seemed to be burning with more power, and the top of the hole seemed to be widening. With more than two hundred men now thrown into the battle, the cost was estimated to be $35,000 a day, including $25,000 for the lost gas and the rest for men's wages and equipment.

The *Bakersfield Morning Echo* reported, "The large tongue of flame has caused curious coloring around the mouth of the well.

The ground has turned peculiar shades of red and gray. Some state that it is chemical action caused by the awful continuous blast of heat and others that the heat alone has caused change on the earth."

That night the steam from sixteen boilers was hurled at the monster. Steam proved ineffective. The fire appeared to be increasing in volume. Standard officials said it was burning lower in the well, which would cause more difficulty in extinguishing it. The call went out for more boilers, and a new approach. Someone suggested the company try chemicals. A baggage car full of carbon tetrachloride was hastily dispatched by rail from San Francisco. The car was transferred at Bakersfield and moved by special train to Taft.

On the Sunday that marked the beginning of the second week of the fire, Standard was ready to launch another attack. Wool-clad workmen hit the stubborn flames with steam from twenty boilers. Twelve pumps mixed fifteen tons of carbon tetrachloride into the steam. Flames sank low into the crater, then with a roar shot high into the air. The battle was not yet over.

By this time the flames that lit the night sky had begun to lose appeal as a novel oil field spectacle and were beginning to assume proportions of a menace. There was some thought that the whole earth in the vicinity of the well might cave in, leaving a vast and yawning cavity where gas had been contained. The fear was also expressed that the well would deplete the supply of gas for the towns of California, plunging them back to the days when coal was used to heat homes. It was not a happy prospect.

It was plainly time to call an expert. Standard turned to Ford Alexander, an oil field dynamiter who only recently had returned to Taft from the boom at Ranger, Texas. Alexander directed workmen in tunneling as near to the mouth of the fire as possible. On Monday, the ninth day, charges of dynamite were placed in the tunnels. The dynamite was set off by electricity. The explosion failed to close off the mouth of the well, though it did alter the hole. Flames continued to leap into the sky, though now they were only half as high. A rumor swept the fields that government officials had visited the well and that the government planned to take a hand. Standard denied the rumor and said its men would continue in charge.

Ford Alexander lost no time preparing another attack. It appeared that dynamite, to be effective, would have to be discharged not from a tunnel but from directly over the hole. To get the dynamite in place, Alexander devised a clever scheme. A cable was strung

Recreation hall on Standard Oil Company's Tupman lease, Elk Hills, 1920s. (Photo from West Kern Oil Museum)

from a wooden derrick some distance from the wild well and anchored across the draw to a post implanted in the ground. A trolley was fitted to the cable so that the trolley could be drawn into position over the hole. Dynamite was attached to the trolley, and the first attempt to work it into position was made that very night, the same Monday that the tunnel charges had proved ineffective. Wheels of the trolley tangled in the cable. The car was hastily withdrawn, its load of dynamite still intact. Alexander decided to wait for daylight for the next try. A decision was made to hit the gasser simultaneously with the dynamite blast and with all the steam and carbon tetrachloride twenty boilers could muster.

Shortly after ten o'clock on Tuesday, August 5, the tenth day, more than one hundred workmen manned positions at boilers, hoses, pumps and cable. Dynamite was skillfully brought into position and detonated, even as steam and chemicals poured into the crater. The dynamite blast cut off the pyre of gas. Flames burst into the air, and the well was quiet. Cheers broke from the workmen. A moment later rocks and dirt blasted from the crater, but the accompanying gas did not catch fire. Men immediately began clearing the ground to begin the effort to cap the well.

Sixteen days later on Friday night, August 21, workmen had reached the casing and dug a large enough hole around it to position the equipment to end the uncontrolled flow of gas. Twenty-six days after it had begun, the flow of gas from Hay No. 7 ended. Standard officials estimated the rate of flow at the time the well was capped at 140 million cubic feet per day. Gate valves were installed and pipe lines laid to put the gas well on production.

Standard's troubles were not over. A series of lawsuits followed in which the federal government contended the company's title to the school section was not valid. In 1921, the Secretary of the Interior upheld the company's title, and it remained uncontested until 1924, when the United States Senate directed the Department of the Interior to again institute suit for the recovery of the land. Finally in 1926, the Supreme Court of the District of Columbia substantiated the oil company's title to the section. By this time, Hay No. 7 had been on production for some seven years. The well had produced an unprecedented forty-three billion cubic feet of gas. It was hailed as the world's most productive gas well. Thirteen years later, renewed litigation resulted in the federal government regaining possession of the school section.

Warning signs advised outsiders they were entering the oil field strike zone. (Photo from Phil Witte)

Striking members of Taft Local No. 6 gathered on a street corner in Taft to discuss strategy. (Photo from Kern County Museum)

On Strike

The purpose of the public barbecue at Buchanan's Pavilion on Wednesday night, July 27, 1921 was to raise funds for construction of a headquarters building by Taft Local No. 6 of the International Association of Oil Fields, Gas Well & Refinery Workers on four lots the union had purchased on Center Street near Sixth Street. The union had had plans drawn by an architect for a reinforced concrete structure two stories high. Cost was estimated in excess of $50,000. Work was scheduled to begin about October 1 with completion as soon afterward as possible. Because of the desirable location, the union planned to rent the lower floor as a source of revenue, maintaining the second floor for meetings and other union activities.

Along with beef barbecued by Dan Marcelas under the tutelage of S. M. Coker, the evening featured a country store where the wheel of fortune spun to select lucky winners, a ball pitching booth where those with good throwing arms might win kewpie dolls and a ring throw where one might "rope" a duck for some future dinner. Later there was a dance.

The mood was relaxed, but there was underlying tension. Men were working under an agreement that would run out in barely another month. For the preceding four years through the period following the country's involvement in the war in Europe, both workers and operators had been covered by an agreement entered into by the federal government and the operators establishing stable wages to insure the nation a continuing supply of oil necessary to the war effort. The agreement was scheduled to expire at the end of August.

Though oilworkers were not aware of it, a decision had been reached six days before the barbecue and dance at Buchanan's Pavilion that directly concerned them. At a conference in Los Angeles, operators affiliated with the Chamber of Mines & Oil had agreed to reduce wages by one dollar a day effective September 1.

The announcement of the proposed wage cut came on August 2, six days after the oilworkers' fund-raising party. The operators' statement said:

"Notices are being distributed by the California Chamber of Mines & Oil to all the oil companies, pipe line companies, gasoline plants and refineries in California, advising the companies as to the action taken at a recent meeting of the oil operators in Los Angeles.

"At this meeting it was unanimously decided that no further conferences be held with the President's mediation commission with a view to the renewal in any form of the existing memorandum of terms which expires on August 31, 1921.

"This memorandum of terms is an agreement between the President's mediation commission and the operators of California covering the working conditions and the wage schedules in the oil industry. It is not an agreement with representatives of the men or unions. It was made a part of the agreement that open shop conditions should prevail in the oil industry."

The statement, after saying a reduction in wages was being made because of decreased cost of living, went on to spell out the new wage schedule.

"The various classifications carry wages of $5.00 for roustabouts; $5.25 for teamsters, light truck drivers; $5.50 for pumpers, stablemen, light tractor drivers, and some machinists' helpers; $5.75 for boiler washers, second engineers, six-horse teamsters; $6.00 for heavy truck drivers, some skilled mechanics, first engineers; $6.50 for workers around the derricks, and other classifications up to as high as $8.25 for heavy fire blacksmiths, and $9.00 for drillers."

The $1-a-day wage cut represented in the case of roustabouts a 16.7 per cent reduction.

A. F. L. Bell, chairman of the operators' committee, said, "The existing agreement with the workers was a war measure, and the war has ended." When asked about the possibility of meeting with federal mediators to discuss the wage cut, Bell stated, "There is no reason why we should meet with the federal mediators as there is nothing to mediate."

Four days later, more than 1,200 members of Taft Local No. 6 voted unanimously to strike on September 1 if oil operators refused to enter into a conference with workers and the government to discuss wages. More than 500 tour men whose work schedules prevented them from attending the first meeting met the following day and gave the same unanimous support to the decision to strike.

A committee sent a telegram to Washington stating that oilworkers had come to believe themselves part of the plan put into effect during the war to insure industrial peace and that the announcement by operators that they would refuse to meet either with workers or government officials came as a "bolt out of a clear sky." The telegram closed with the warning that unless operators could be brought into conference, "the oil industry will be shut down immediately after September 1st." The telegram was sent to President Warren G. Harding, Secretary of Labor James J. Davis, Secretary of Commerce Herbert Hoover and Secretary of the Navy Edwin Denby.

In Oilworkers Hall on Main Street in Taft, wives, daughters and sisters of union members met to organize a women's auxiliary. The auxiliary had been in planning stages for some time, but men had favored postponing its organization until its members could be given quarters in the new building to be erected on Center Street. With construction of the building now delayed by what a union spokesman called "the unsettled condition of the industry at present," women took matters into their own hands, insisting on a call for a mass meeting. The hall was packed. No men were allowed inside until 10:30 P.M.—three and one-half hours after the meeting began—when the women pronounced their organization complete with Mrs. Grover Cain elected president; Mrs. Helen Miller, first vice president; Mrs. John Hurley, second vice president; Mrs. James Peel, recording secretary; Mrs. Paul Tooker, treasurer; Mrs. M. C. Starkey, guard; and Mesdames J. H. Thrasher, Charles S. Robertson and W. S. Jan Dell, trustees. To Taft and Bakersfield newspapermen waiting in the main office of the hall, the women sent out the message, "The men fought for us during the war and we will fight for them now."

In Bakersfield, some 400 members of Local No. 19, including many women, paraded through the streets handing out flyers that stated:

"We are going to strike September 1st if the oil operators still refuse the request of the United States government to enter a conference to continue the government plan which has kept industrial peace in the oil fields for the past four years. We helped win the war for Democracy when the kaiser defied the U.S. government—we helped lick him. We will not stand for our government being defied by an autocracy at home."

Four days before the strike deadline, Taft Local No. 6 sent out a call for ex-servicemen in the union ranks to meet at union headquarters.

The *Daily Midway Driller* reported, "When this organization puts out a call for ex-servicemen, it's like calling the roll, for there is no other industry that was better represented in the world war than that of the oilworkers. It will be remembered that during the early days of our entry in the struggle the papers all over the country carried stories in regard to the flocking to the places of enlistment by the oilworkers from all over the fields and how the oil wells were about to be closed on account of the great numbers of the workers who had joined the colors."

Two days before the strike was to begin, John A. Burns of the Taft local announced that Shell Oil Company had signed an agreement under which the company would continue in force the working terms of the agreement of the past four years and the oilworkers would accept a reduction in wages of $1 a day. Burns said that twenty-five smaller companies had agreed to meet with the federal mediation board and oilworkers for discussions expected to lead to a similar agreement. He also announced that ex-servicemen in the union had been organized into the Law and Order Committee to protect property should a strike develop.

News of the Shell agreement coincided with receipt by the union of a telegram signed by Charles T. Connell, commissioner of conciliation of the U.S. Department of Labor, and E. P. Marsh, formerly of the President's mediation committee, who made up the two-member Federal Oil Board. The telegram requested that the strike order be revoked, saying, "The Federal Oil Board believes that if both the workmen and employers exercise restraint and cool judgement the situation so pregnant with unpleasant possibilities may be eventually composed to the satisfaction of all concerned."

At a hastily called meeting in Taft, oilworkers voted to rescind the strike vote. Union officials said they would continue the effort

Daily Midway Driller, September 15, 1921. (Beale Memorial Library)

Members of Fellows Local No. 13 gathered in front of union hall to await assignments manning roadblocks. (Photo from Phil Witte)

to bring about a conference with operators and federal authorities on adjustments in the announced wage cut and on terms of the agreement in existence through the past four years.

The West Side Merchants Protective Association, though recognizing it was contrary to its by-laws to take part in labor or political activity, sent a telegram to the Department of Labor recounting operators' refusal to meet with representatives of the government and oilworkers and declaring, "We the directors of the West Side Merchants Protective Association of Taft go on record as disapproving the attitude of a part of the operators and their efforts to prevent a conference."

On Monday night, September 5, union members celebrated Labor Day in Taft with a 6:00 P.M. parade that formed at Standard Oil Company's Section 14 property on the west end of the city and marched the length of Center Street, doubling back on Main. The oilworkers' union and members of the auxiliary headed the parade, led by Secretary Val Walters and General Organizer Burns. Pete Duff led the eight-piece band.

Oil operators steadfastly refused to meet with government officials or union representatives. In Los Angeles, *California Oil World* reported that since the first of the year, production in California had exceeded consumption by seven million barrels. "If there is to be a strike," the weekly trade paper said, "now is a good time for the operators."

On Sunday night, September 11, oilworkers met in Taft and voted to strike. Other locals in Bakersfield, Coalinga, Fellows, Lost Hills, Maricopa and McKittrick also voted to strike. The walkout in West Side fields began Monday.

One of the first actions of the Taft local was to send a delegation from the Law and Order Committee to visit those who had liquor to sell oilworkers. Though prohibition was the law of the land, hard liquor was readily available. One supplier at McKittrick offered bonded Old Crow and Old Yellowstone for $1.50 a quart, removing the whiskey from counter display only when the constable happened to be on the premises. The union let it be known it would tolerate no drunkenness. A spokesman said, "If any of the boys on strike secure a drink of liquor, the Law and Order Committee will find out where it was secured and the place will be raided by police operatives."

Though gambling, too, was illegal, card games normally ran twenty-four hours a day in various establishments in Taft. The strikers' delegation visited gaming places, requesting that games be closed down at six o'clock each evening for the duration of the strike. Gamblers agreed to do so. The committee made plain that the aim and desire of the union was to preserve quiet and not permit any damage to the property of the various companies through an act of violence for which the union, though innocent, would receive full condemnation.

The Taft local posted a list of companies and individuals who had signed up with the union and announced workers would return to their jobs. The list included Lake View No. 2 Oil Company, Elk Horn, Bairstow, Spreckels, Obispo, Brookshire, Millie Frances, R. E. Graham, Midway Northern, Brooks, Taft City Annex, Olea Fuel Company, Anthony Oil Company, Browngold Oil Company and Stockton Midway. Acknowledging that all were small and employed only a few men, union officials said they felt confident that the action would pave the way for future progress in the effort to secure the signatures of larger companies.

The walkout did not extend to Standard Oil Company. The company, which did not recognize the union, had proposed a ten per cent wage cut to become effective October 1. The reduction was less than the across-the-boards $1-a-day cut agreed on by other operators and meant that after October 1, Standard would be paying higher wages than others.

Strikers and sympathizers joined in a big parade through the streets of Taft on Tuesday night, September 13, disbanding at the Sunshine Theater, where Walter Thompson Mills, of Berkeley, spoke on "The Victories and Defeats of Organized Labor." Mills, a noted lecturer, had been the first representative of the American Federation of Labor to the British Trades Congress.

Late that night, word reached union headquarters that a special train was preparing to leave Bakersfield for Taft carrying nearly 250 men described as strikebreakers recruited from the San Francisco Bay area.

Strikers immediately began to mobilize. One member of Fellows Local No. 13, arriving back in town in the small hours of the morning after attending a dance in Bakersfield, was astonished to find the main street of the oil community ablaze with lights and swarming with excited men and Model T Fords. At union hall, he was told, "Go home and get your gun. We're going to stop a train." He joined others armed with shotguns and rifles in a procession of Model T's wending their way toward Pentland Junction just outside Maricopa.

As dawn was breaking, more than 1,000 oilworkers from West Side fields met the special train as it pulled into the junction. The group included many women. Strikers stopped the train, and an armed delegation went aboard to confer with spokesmen for the men inside. An agreement was quickly reached to the effect that the train would go back to Bakersfield and that the men aboard would not return. The conductor refused to turn the train around without orders from Bakersfield. During the wait, an occasional passenger lifted a window shade to see what was going on outside. Each look was met with a barrage of derogatory remarks, and the shade was hurriedly lowered. After orders came, a committee of one hundred oilworkers boarded the train and accompanied it back to Bakersfield to make sure that those aboard carried out their promise not to return to the oil fields. No shots were fired, nor were any blows struck.

Later in the day, Commander H. B. LaMonte of Taft's American Legion post addressed himself to the matter of a request by the union's Law and Order Committee that it be allowed to use the Legion hall. Strikers had been evicted from company bunkhouses, and many lacked a place to sleep.

LaMonte said after consulting with members of the post's executive committee he was convinced "it would be right and just and proper to throw the hall open for use of the Law and Order Committee." He added, "The American Legion, standing as it does for law and order, and for the law of this government, which is founded upon reason, law and order in all things, felt that in complying with the request of the committee for the use of the hall it is not taking sides one way or the other, but is occupying a useful, neutral position, and trust that the strikers, operators and general public will consider that the Stanley H. Little Post No. 70, American Legion, is acting in all fairness in granting the request of the Law and Order Committee that ex-servicemen and American Legion men serving upon the Law and Order Committee be domiciled in the hall during the present disturbance."

Strikers called an open air mass meeting of all organized labor in Taft and vicinity on the property of the Pioneer Mercantile Company at the corner of Fifth and Main Streets that night. Afterward there was a public dance at Buchanan's Pavilion. L. R. Buchanan donated the pavilion and music. The women's auxiliary served sandwiches and hot coffee. All proceeds went to the strike fund.

On the following day, while strikers kept a close watch to see that no strikebreakers were brought into the fields, Walter J. Yarrow, who had come from district offices in San Francisco to direct the strike, said men would return to work if operators would agree to meet with the government's mediation board. When asked what strikers wanted, he said they merely wanted to know where they stood relative to any further wage cuts.

It was rumored that the government would step in to operate wells that might be damaged by being shut down on properties held by operators under lease from the government if the operators were unable to work the wells.

A rumor swept the fields that carloads of strikebreakers from Southern California were enroute to Taft.

The Law and Order Committee said open season had been declared on bootleggers and many had moved on.

At the regular Thursday night meeting of the West Side Merchants Protective Association at Smith Brothers Hall, the association dispensed with regular business to listen to a plea from William Pope, president of Taft Local No. 6, asking that the local be allowed to solicit members of the association for funds. The association granted the request. Businessmen subscribed $4,000 to the strikers' fund.

On Friday, operators met in San Francisco's Palace Hotel to discuss the strike, which now involved some 8,000 oilworkers in San Joaquin Valley fields. They formed an organization to deal with the situation, electing as officers M. H. Whittier, president; A. F. L. Bell and D. S. Ewing, vice presidents; L. P. St. Clair, E. W. Clark, Paul Shoup, A. C. Diericx, Allard D'Heur, Lionel Barneson and C. J. Berry, executive committee. George M. Swindell was elected secretary. They named their organization the Oil Producers Association of California, and declared that it represented between 80 and 90 per cent of the industry affected by the strike.

The association issued a statement saying that it expected the constituted officers of the State of California to see promptly that rights of persons and property in the oil districts were maintained.

Charging striking oilworkers with misrepresentation of facts, the association said there was no dispute about wages or working conditions and that the purpose of the men heading the strike was to bring pressure by force upon the government to interfere and assume obligations of mediation and arbitration "which they intend shall develop into control of the industry."

Stating that oil producers believed the slogan "more business in government and less government in business" to be a proper one, the association further charged:

"At this time a lawless condition exists in the oil fields of the San Joaquin Valley. The highways into those fields are guarded by strikers, vehicles using them are stopped and passengers permitted to use them only by permission of strikers. The constables in these oil districts have deputized several hundred strikers, thus enabling them to carry weapons openly. At night barricades have been erected across the highways. Men who attempted to remain at work, except at a few field transportation pumping stations have been obliged to quit. Guards sent in to protect the companies' properties against depredations have been turned back. A special trainload of such guards, sent not to operate the properties, but merely to protect them, was derailed on the Sunset-Western railway, taken possession of by the strikers, and forced to return to Bakersfield. Officers of the companies have been shot at. Stating it baldly, the properties of the companies have been taken possession of by the strikers. In some of the towns in that section, no merchandise, not even drugs, can be bought except under permit from the strikers, and in some instances, union employers are not allowed to leave town except with their permission.

"To yield to their demands under such conditions is to yield to a force that ignores the law while claiming to have respect for government."

The association charged that if the demand by union officials for government mediation were applied to industry generally in the United States, it would inevitably mean socialization. "It is inconceivable that a self-respecting commonwealth should apathetically yield to the super-government set up by certain strikers in the oil districts, riding over the rights of both men and property guaranteed by our government."

In Taft, Local No. 6 quickly denied that oilworkers had taken the law into their hands, were interfering with legitimate business in the strike area, or that an operator had been fired upon.

The union's statement said:

"Members of the Law and Order Committee of the Oilworkers have perfected an organization throughout the entire district, and our men are guarding every approach to the fields with a view only of preventing strikebreakers from entering the fields. A campaign is also under way against the bringing in of 'bootleg' whiskey, intoxicating drinks or guns and ammunition to the fields at this time, the union men feeling that the present time is one which calls for careful judgment on the part of every member.

"Although many cars have been stopped by pickets everyone is given courteous treatment, and in no case has anyone been discom-

NOTICE

The Cooks, Waiters and Waitresses Union of Taft wish to state that the rumors circulated that the Shamrock Restaurant is unfair to organized labor is untrue, and that the Shamrock Restaurant is fully entitled to the hearty support of all organized labor.

(Signed) A. Varley,
Pres. Local No. 771.

NOTICE

Taft, Sept. 19, 1921.

This is to advise that the circulated report of The California Market Company being unfair to organized Labor is untrue. This Market is complying in every way with their agreement, and are therefore entitled to the support of organized labor.

The aboe is to refute all reports to the contrary.

Butchers' Local No. 652,
By C. B. Stapp, President.

Daily Midway Driller, September 19, 1921. (Beale Memorial Library)

moded where their mission was not found to interfere with the strikers in their plan to keep the fields free from liquor and strike-breakers.

"The statement that strikers would not permit the purchase of drugs or merchandise without their permission is absolutely false in every detail, but the oilworkers have refused to sanction the sale of intoxicating drinks and will continue to do so.

"We have accepted a dollar a day reduction. We have accepted the open shop. All we ask is that the operators sign a similar agreement arranged between the Dutch Shell and Doheny interests and other companies with the government. When that is done we can return to work."

At a well-attended Sunday evening service of the Taft Baptist Church, members unanimously adopted a resolution which stated:

"We commend the fairness and orderly manner in which the strike is being conducted, and we desire to assure them (the strikers) that so long as the strike is conducted in a lawful and orderly way they will have our prayers and our co-operation in their struggle for justice."

The Loyal Order of Moose extended moral as well as material aid. Members voted to donate hot dogs, rolls and coffee to the makeshift kitchen adjoining union headquarters where hundreds of men were being fed daily.

Members of the American Legion voted unanimously to endorse the action of the executive committee in throwing open the Legion hall for the benefit of those serving on the Law and Order Committee. The move was said to have been done not with the view of taking sides in the labor controversy but because members felt that while their buddies were out of work, some of them without funds and needing a place to sleep, they should have the assistance of the post to its fullest extent. Chairman Conklin of the Legion's boxing committee reported on plans to sponsor amateur boxing bouts in Taft and said it had been decided that proceeds from the first card consisting of five four-round matches would be donated to the striking oilworkers.

The Women's Auxiliary of Stanley H. Little Post No. 70 announced that the social planned for the following Thursday night, September 22, was being postponed for the present due to the fact that the Law and Order Committee was using the Red Cross rooms in the Legion hall.

On Wednesday, September 21, ten days after the strike had begun, a three-man committee from the Oil Producers Association of California arrived to meet with Kern County Sheriff Boone Newell and to look over the situation in the fields. The three men were D. S. Ewing, Allard D'Heur and A. F. L. Bell. They declined to say what action operators might take but said any suggestion that they were planning to yield to the strikers was without foundation. They also took exception to the opinion expressed in some quarters that idled wells might be ruined through encroaching water, saying they were familiar with problems of water control and would take steps to prevent damage.

A strikers' roadblock netted a well known political figure in the person of A. J. Wallace, former lieutenant governor of California. Wallace, a director of Transport Oil Company, said he was questioned, but treated "with every respect."

The West Side Merchants Protective Association sent a wire to President Warren G. Harding, California Governor William D. Stephens, Kern County District Attorney Jesse R. Dorsey and Kern County Sheriff Newell stating that the association had noted "with considerable surprise" articles in various newspapers alleging cases of lawlessness on the part of striking oilworkers. "No acts of lawlessness have come to our observation," the association's telegram said. "On the contrary, we commend in the highest possible terms the manner in which this strike has been conducted by officials of the Oil Workers union together with the Law and Order Committee."

In Maricopa, merchants issued a statement supporting the strikers. "All the oilworkers are after," the statement said, "is a square deal for all."

Two days after the committee from the Oil Producers Association of California had visited Kern County, M. H. Whittier, association president, charged that striking oilworkers had turned the West Side into a veritable "little Russia." Whittier said producers were not "in a mood to make overtures of any sort." He said they would let matters take their course, feeling that when strikers realized the true situation, they would return to work voluntarily. "We will not pull any West Virginia stuff," Whittier said, "and would rather shut down entirely for six months than to take such action."

Martin G. Madsen, secretary to Governor Stephens, met with strike leaders in District Attorney Dorsey's office in Bakersfield and told them with their roadblocks they were attempting to take the law out of the hands of legally constituted authorities. Strikers protested that they were using peaceable means. Madsen replied that when a man appeared in the road with a club in "a menacing attitude," he was not using peaceable means and had no right to do so.

Following the meeting, strike leaders issued orders that Law and Order patrols throughout the strike zone cease stopping highway traffic until further notice.

Amid grumblings of a sellout, Walter J. Yarrow extended the olive branch to operators in the form of a telegram to A. L. Weil, a former chairman of the oil operators' committee who was a large stockholder and legal counsel for General Petroleum Company. Yarrow appealed to Weil to use his influence to bring the strike to a "quick and satisfactory conclusion" and stated that the oilworkers were "open without further jeopardizing the interests at stake to anything that will bring industrial peace and harmony, with victory to both sides."

While strikers awaited a response, the Emmett Club of Taft announced a benefit card party and dance to be held at St. Mary's hall on Friday evening, September 30, for the striking oilworkers.

It was announced that the ball game scheduled between Taft and Tulare on the local diamond for Sunday afternoon had been called

Bean growers pitched in to help striking San Joaquin Valley oilworkers. (Photo from Phil Witte)

off and that "on account of existing conditions" games would not be played in Taft for the present.

Oil operators filed more than one hundred suits in Kern County Superior Court to force the eviction of strikers and their families from lease houses.

Pay call sounded late Sunday afternoon, September 25, for strikers who had been employed by Pacific Oil Company. They lined up on Center Street and filed by the company pay clerk's car to receive wages for the ten days they had worked during the month prior to the time the strike was called. A number of men received time vouchers, indicating the company was through with their services.

On Tuesday, September 27, the Oil Producers Association of California made its reply to the overture from strikers for "peace without victory." Producers refused to have the federal mediation board a party to any wage agreement on grounds it would mean "an attempt to nationalize the industry" and would encourage "deliberate shirking of work."

In a statement signed by M. H. Whittier, president, and G. M. Swindell, secretary, the producers' organization said it had adopted as its cardinal principle "the right and the obligation of each employer and of his employees to agree on their relations without government control."

In a circular addressed to members, the association's officers said that for the present no attempt would be made to operate properties affected by the strike "except where there are sufficient men who can be amply protected." The circular went on to say, "Notwithstanding statements to the contrary, so far as the association knows, no attempt has been made by the operators in the district affected by strike conditions to take in strikebreakers or armed guards."

In Taft, the *Daily Midway Driller* reported the consensus of opinion among oilworkers "as nearly as can be gathered" pointed to strong opposition to anything which would appear to nationalize the oil fields. "Many of the workers," the newspaper reported, "while strongly in favor of a contract with the operators insuring against a further cut within a year, are nevertheless willing that the mediation board members for the government should be left entirely out of the dealing. Some of the strikers, however, advanced the thought that the signature of the government on the contracts of wages and working conditions would merely certify that they will see fair play between both parties."

A notice in the newspaper advised all oilworkers on strike to register daily with the union and asked that all in need notify the Relief Committee in newly-established quarters in the Fox Hotel building. The notice asked citizens who had sleeping quarters for men or cots to loan to report to the Relief Committee and requested that those who had old clothes to donate, particularly clothing that could be made over for children, to please notify the committee.

A rumor circulated that the volume of material being shipped to fields in Southern California which were not affected by the strike indicated operators planned to increase production from that district to overcome any shortage resulting from shutdown of wells in the San Joaquin Valley.

There were disquieting reports that automobiles bearing four or five men were entering the fields on various roads and later returning occupied only by the drivers.

Governor William D. Stephens issued a statement in which he declared that the sheriff and district attorney had ample authority to maintain law and order. His personal secretary, Martin C. Madsen, said it was essential that the strike be ended as quickly as possible and denied that he was in any fashion attempting to enlist the Federal Oil Board in any endeavor to settle the strike.

As the third week of the strike drew to a close, quiet prevailed, broken only by recurring rumors that strikebreakers were on the way to West Side fields.

Walter J. Yarrow of the oilworkers' union sent another telegram to A. L. Weil. Referring to the charge of inefficiency leveled by the producers' association, Yarrow said, "We will concede some inefficiency in instances where those in charge of operations bullyrag, intimidate and threaten the workers." With regard to the producers' claim that the government's underwriting of agreements was acceptable only during war, Yarrow said, "It would look like good business for employers, employees and the public if the signature of the government upon an agreement between oilworkers and operators would end the strike."

In San Francisco, G. M. Swindell, secretary of the operators' group, issued a statement declaring that operators were "in the fight to a finish" and added that there would be no deviation from the principle that they would refuse to make government mediators a party to any agreement with oil field workers.

Swindell said reports reaching operators indicated that in the Midway district strikers were frightening the wives and members of families of loyal employees and that "mop-up gangs" were threatening and intimidating others who remained at work and had "beaten a cripple and a youth acting as a messenger."

Swindell added, "With the general business depression and consequent sub-normal demand for petroleum products, it is doubtful whether there will be nearly as many jobs available as there were when the strike was called."

Swindell also charged that Industrial Workers of the World literature was being distributed in the oil fields.

As the fourth week of the strike began, Secretary of Labor Davis sent a telegram to the Oil Producers Association of California asking the association to send a delegation to Washington in an effort to arrive at some solution. "We have no desire to name a third party as referee," the message said, "and there is no quicker way to take the government out of a controversy than for both sides to show they can agree between themselves."

On the same day in Taft, the first strike benefit checks were handed out. The checks were $10 for single men, $15 for married men. Union Secretary Walters said that benefit checks would be paid again in another week. The checks brightened business in Taft, and some predicted the strike would end in another ten days or two weeks.

California Oil World reported that more than 1,000 special deputy constables had been appointed since the strike began, and that practically all were strikers. The publication said that Kern County Sheriff Newell had deputized some foremen and superintendents, and added that "ever since the strike, it has been difficult for those unfriendly to the strikers to get service, accommodations or meals in the oil field towns."

While strikers waited to see what response operators would make to the Secretary of Labor's request for a conference in Washington, Paul Shoup, president of Associated Oil Company, Pacific Oil Company and affiliated companies, said of the strike:

"This is not a case of a single acute dispute involving need for arbitration. It is a deliberate attempt to maintain perpetually a government board now called a federal board of government employees to have the final voice in determining wages and working conditions in the oil industry.

"If this be a good thing, why not apply it to the oil industry as a whole? Why not apply it to all industry in the United States?

"Such a course is unthinkable and directly contrary to the principles of our American form of government.

"The oil industry, like all other industry, should be able to get on by full and frank discussions, man-to-man consideration, between employers and employed. The oil companies in the Producers'

Strikers thronged the main street of Fellows, eagerly awaiting news of the attempt to bring oil operators to the conference table. (Photo from Phil Witte)

Association are firmly of the conclusion that it will be a great handicap to the industry and the capital and labor employed therein to establish this outside control.

"The industry in this state when prosperous has used that prosperity to enlarge; to prospect for new fields to provide more and more employment. You who are old-timers compare today with twenty years ago. The stockholders have as a whole received only modest returns. The Associated Oil Company has never paid more than 6 per cent per annum dividends to its stockholders. All the other earnings, when any, have gone back into the properties and provided more jobs. And they believe that the assumption by anybody outside of the industry of the power to decide questions of vital nature will hurt the industry and the constructive program heretofore effective."

In Kern County oil fields, members of women's auxiliaries visited various leases in an attempt to persuade non-union men to leave their jobs.

In Taft, the post office said in the three weeks since the strike had begun nearly 2,000 persons had changed their addresses, the greater number now receiving their mail at the general delivery window. The *Driller* said commercial establishments reported a falling off of business by perhaps as much as twenty-five per cent. The newspaper said streets no longer seemed crowded and that many seemed to have left town.

The American Legion at its regular Friday night meeting said it was dropping plans to hold amateur bouts in Taft "until such time as conditions might justify the venture."

Union officials explained the fact that fewer pickets were to be seen by saying pickets had achieved a greater efficiency. Officials claimed that morale was excellent and said that reports of many strikers leaving the area were exaggerated and that only a few "weak sisters" had departed. They said that in the most productive part of the entire West Side fields one company on strike was having trouble with caving wells and pyramiding water while its neighbor, not on strike, had more than doubled production.

Oilworkers held a mass meeting on Sunday, October 9, at the Sunshine Theater at which International President R. H. Stickel spoke. Later in the evening a parade was held after which an open air meeting followed on the street adjoining union headquarters. Mrs. Tom Reardon told strikers they could not consider they had gone through a strike until they had met some of the troubles which had come to striking West Virginia coal miners, nor could they feel they were really striking when it was shown that the garment workers had been on strike for fifty-two weeks before they had won their point.

As the second month of the strike began, Strike Advisor Yarrow expressed confidence that the government would act to end the strike.

On Thursday, October 13, strikers wired Governor Stephens asking for troops to prevent bloodshed following a clash at Kern River between the union's Law and Order Committee and "gunmen of the Associated Oil Company." T. W. Powell, captain of the Law and Order Committee, said, "The gunmen have challenged us and we will accept the challenge."

Yarrow said deputized gunmen of the Associated had challenged deputized strikers charged with keeping the peace. As a truckload of armed guards had gone past a Law and Order outpost, Yarrow said, George Burton, a former Bakersfield policeman now employed by Associated, was said to have remarked, "I am going through your outposts, and the first man who steps in my way I am going to drop him. I never miss my shot."

In Taft, a Law and Order patrol working with T. J. Nicely, federal prohibition agent, and Deputy Sheriff Dupes of Bakersfield discovered a large still at Carrizo Plains and arrested the operator.

In mid-October, production figures were released showing that output during September had declined in the Midway-Sunset field from a pre-strike figure of 133,679 b/d to 98,013 b/d and in the McKittrick field from a pre-strike 6,989 b/d to 2,943 b/d, or a total drop of slightly less than 40,000 b/d.

From Washington came word that Thomas O'Donnell of Los Angeles, president of the American Petroleum Institute, had met with Secretary of Labor Davis to present the oil operators' view of the strike. O'Donnell was reported to have urged that the government adopt a "hands off" policy. Afterward, O'Donnell called on President Harding.

In California, the Oil Producers Association of California declared that as a matter of employment, a man's affiliation with a labor union depended entirely upon the interpretations he placed upon his obligations to the union; "first, should they be opposed to his superior obligations requiring as a personal duty that he do a fair day's work for a fair day's pay; second, how far these obligations to his union embarrass or hinder the existence of the family relations which should exist between managers and employees under free conferences."

On Tuesday, November 1, eight weeks after the strike had begun, strikers began voting on the question of whether they should end the strike. The district council recommended they vote to end it. The reason, Strike Advisor Yarrow said, was to heed a telegram from Secretary of Labor Davis requesting that the strike be called off "without prejudice." In the telegram, Davis called for "the resumption of production in this great basic industry which is so vital to the nation."

Before going to their posts that night, members of the oilworkers union gave a charivari in Taft for Sam Lewis and his bride, filling more than twenty cars with oilworkers and their wives and sweethearts. The caravan went to the home of the young couple in South Taft, placed them in a truck at the head of the parade and proceeded through town, stopping in front of the union hall, where Lewis made a speech. Afterward, Lewis and his wife entertained at their home, serving refreshments. At midnight, men departed for their various posts of picket duty, carrying with them the memory of a pleasant evening.

On Wednesday, November 2, G. M. Swindell of the producers' association, issued the following statement:

"The decision of the district council of oilworkers recommending to the various locals that the strike be called off will have no appreciable effect upon the plans of the Oil Producers Association and member companies.

"We have already issued our declaration of principles and defined the procedure that will be followed by member companies when operations are resumed in the oilfields affected by the strike.

"The process of individual selection in employment will be used and member companies will treat directly with their own employees. No interference between employer and employee by any outside agency will be tolerated."

On Thursday, November 3, Walter J. Yarrow wired Washington that men had voted to return to work.

The strike ended against a background of complaints of wholesale firing of men who had struck, an attempt to burn bridges on Highway 119 linking the West Side oil fields with Bakersfield, sporadic shooting and the alleged flight of Strike Advisor Yarrow to Mexico with union funds.

Hooded Terror

The handbills that appeared on the West Side in early February 1922 bore the heading Ku Klux Klan. The message said:

"First, last, and only warning to gamblers, gunmen, bootleggers, loafers, lawbreakers of every class and description, black, white or yellow. We are sworn to preserve the sanctity of the home, the virtue of our wives, mothers and daughters, and we mean to do just that. We stand squarely back of law and order. We demand strict, impartial enforcement of the law. We are strong enough to enforce these demands, and will back our officers to the last man in enforcing the law. We demand that the town of Taft and the county of Kern be made clean, and that happiness and welfare be safeguarded. No law-abiding citizen need fear our coming, but he who defies the law and common decency will do well to change his course. Lawbreakers, you cannot escape us. We know who you are, what you are, and where you are. Change your ways this hour lest you be stricken as with lightning from the sky. The good will welcome us: the evil will meet with a swift and stern retribution. We have given fair warning. Beware!"

Soon afterward an ominous parade moved boldly down the streets of Taft. The nighttime procession consisted of eight automobiles with lights dimmed and license plates covered. Occupants wore the peaked hoods and white robes of the Ku Klux Klan. From the cars, muzzles of rifles pointed menacingly at silent spectators.

Though there were rumors of warnings to various individuals and of beatings for some, the appearance of the Ku Klux Klan on the West Side brought no official reaction in Taft from Porter Munsey, who was the city's marshal, or Stanley Abel, who as supervisor of the Fourth Road District was the highest elected official on the West Side. Abel, chairman of the Kern County Board of Supervisors and also secretary of the County Supervisors Association of California, had only recently returned from the American Roadbuilders convention in Chicago, stopping long enough in Phoenix, Arizona on the return trip to make favorable comments about roads in the

Stanley Abel was supervisor of the
Fourth Road District when the Ku
Klux Klan appeared on the West
Side.   (Photo from Kern County
Museum)

state's Maricopa County. His remarks had been printed in the *Arizona Gazette* and reprinted in Taft's *Daily Midway Driller*.

The *Driller* took no immediate editorial notice of the Ku Klux Klan, filling its columns instead with tributes to Abraham Lincoln and George Washington, whose birthdays were celebrated in February, and with local news items of a social nature. In its advertising columns, the newspaper carried ads soliciting investment in out-of-state oil ventures, including one in El Dorado, Arkansas, where an aggressive young promoter named H. L. Hunt was offering to make investors rich through a program he called "guaranteed gushers."

On February 21, Kern County's District Attorney Jesse R. Dorsey confirmed in Bakersfield that his office for the past two weeks had been investigating a complaint lodged by Eli Andrews, a rent car driver in Taft, that he had been beaten by a gang of armed and hooded men.

City Marshal Munsey, when questioned about the incident and its possible relationship to the warning handbills, said efforts to trace the origin of the handbills had failed.

On Saturday, March 4, the *Driller* in its first mention of Ku Klux Klan activities on the West Side reported that the federal government through agents of the post office department had begun an investigation of what was described as "the increasingly bold activities of groups of armed and hooded men who have established a reign of terror throughout the length and breadth of the West Side oil fields of Kern County."

The newspaper said gangs, posing as members of the Ku Klux Klan, had within the past few weeks taken out, beaten and tarred or oiled and feathered three businessmen, forced nearly a score of others to leave the district, and sent warnings to perhaps fifty.

The newspaper reported, "The increasing activity of the night riders reached a climax in the past few days with the disappearance of Eli Andrews, local rent car driver, who is said to have been taken out for a second time and severely dealt with, and with the beating and oil and feathering of George B. Bowman, prominent Maricopa druggist, who was compelled to sell his store and leave the city two weeks after the attack."

Speculation was that Andrews had died of injuries inflicted by the night riders, but the *Driller* reported that a thorough search of hospitals and undertaking establishments had failed to verify the rumor.

The *Driller* said of the Bowman incident, "Bowman, who owned the Gate City drug store, was called to the street by an unmasked stranger, who said that he was wanted at a car outside. As he approached the machine, a touring car with side curtains, the man who had called him out thrust a gun into his back and ordered him into the machine. Three other men inside the car grabbed him, pulled him inside, blindfolded him and threatened to kill him if he cried out. Another carload of men followed.

"After a ride of 20 minutes Bowman was taken from the car with a rope around his neck, still blindfolded and protesting, and a formal charge, said to have been signed by a Coalinga woman, was read. Bowman was asked to confess, but he strongly denied the charge and continued to do so even after he was stripped, beaten to the ground with ropes, covered with oil and feathers, and returned to Maricopa with orders to leave town within two weeks. A physician treated him for serious bruises.

"Bowman, who is well liked throughout the West Side, declared to close friends that he would not leave the city if they killed him, but later, when his wife became hysterical with fear for him, he sold out his business for some $30,000 and departed for Los Angeles Wednesday, the last day of grace."

It was rumored that among others who had left was a driller who had received an anonymous warning accusing him of being an "inveterate card player" and advising him against gambling. The driller, according to friends, had sold his house at a loss of $2,000 and hurriedly departed.

On Sunday night, March 5, the Reverend Van Dyke Todd of the Baptist Tabernacle in Fellows, who several days before was said to

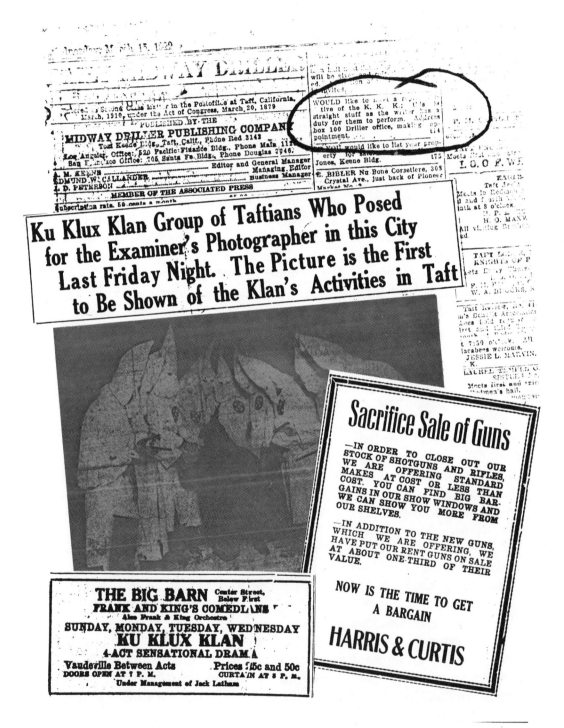

Clippings from the *Daily Midway Driller* with February and March dates, 1922. (Beale Memorial Library, Bakersfield)

have received a letter signed with a black hand warning him to quit preaching against lawlessness, was delivering the evening's sermon when five automobiles carrying approximately twenty-five men dressed in Klan regalia pulled up outside. Without a word, three men who were fully armed stepped from a car, entered the church and walked toward the pulpit. A woman who was present later said, "Most of the women in the church were scared to death and some of the men were white-faced with fear." The leader of the three intruders, who wore a hood with peaked cap and long white gown of starched muslin, told the pastor the Klan was in full accord with his views and had had nothing to do with the warning he reportedly had received. The Klansman made a donation of $15 and the group left the church.

Deputy District Attorney Allan B. Campbell visited the West Side to look into the situation and, according to rumors, found no one willing to talk. Speculation was that fear of the armed, masked bands was so great victims were afraid to go to authorities.

It was reported that John N. Pyles, a private detective with offices in Bakersfield, had been approached by unidentified interests to make a thorough investigation of Klan activities on the West Side. Pyles had formerly lived in Taft, but since the oilworkers' strike in 1921, when he had represented various oil operators, had lived in Bakersfield.

On Wednesday, March 8, the *Driller* disclosed that the first case in which a woman had been run out of town had come to light. Mrs. Mary Barnes, proprietor of a rooming house on Center Street between 1st and 2nd Streets, had moved out after having been warned to close up and leave within twenty-four hours. Neighbors, who declared the house was "questionable," said Mrs. Barnes had admitted that twelve men dressed in the white robes of Klansmen had surrounded the house on the preceding Friday night and demanded that she leave within twenty-four hours. "It won't take me that long," Mrs. Barnes was quoted as saying. She reportedly had gone to Los Angeles.

From Maricopa came word of the formation of an organization described as ready to receive any night riders, whether white-hooded or black-hooded, with buckshot "and ask questions later."

Eli Andrews, who had been thought dead, was reported to have left town after his sister and mother urged him to do so. They said he had not received any second warning and had not been molested after the first assault. The *Driller* reported that Andrews, "painfully bruised, half frozen from exposure, and covered with tar and feathers, lay under the floor of a building here for several hours before he dared to go home. It was learned that he had been charged by Klansmen with bootlegging and selling narcotics to minors before he was beaten to the ground with ropes."

In Bakersfield, Undersheriff Roy Newell issued subpoenas for some twenty individuals who were reported to have received warnings or were directly connected with law enforcement in the West Side oil fields to appear before the Grand Jury for the taking of testimony.

District Attorney Dorsey said he was handicapped in handling the situation by not having evidence of the identity of any of the alleged hooded attackers.

In the Thursday morning, March 9, mail, Editor A. M. Keene of the *Driller* received a letter mailed from Los Angeles the preceding day. The letter said:

"Let this be a warning to you and to all those connected with your office and your correspondence which means those who send out reports from the fields.

"We don't want any more such reports as have been sent out and insist that you take unto yourselves that silence is golden.

"If you do not heed this warning we will be compelled to deal with you as we have dealt with others. K.K.K."

The newspaper published the letter. Late on that same night, another letter was pushed under the door. The letter said:

"We note the so-called K.K.K. letter published in your paper this date. We beg to advise you and all others that receive letters which are supposed to come from us, that unless they are written upon this letterhead, they are not genuine.

"There are no fingerprints on this letter, which you may turn over to Little Pyles or the Fresno Screech, it wots us not.

"Taft Klan No. 3, Ku Klux Klan, Realm of California."

When advised of the letter, Detective Pyles said, "I am getting sick and tired of this whole business. If anyone will put up the necessary money to finance me in this investigation, I will expose the whole outfit from start to finish, name those responsible for the outrages in the fields, and settle once for all in the minds of the public just what is underneath it all."

In the Friday, March 10, issue, the *Driller* published an open letter to the Taft Klan acknowledging receipt of the Klan's letter and stating, "The *Midway Driller* takes this occasion to urge your body through its leaders, to give to the public through this paper, a clear and concise statement relative to your stand and to incidents of the past few months."

The open letter drew no response.

In Bakersfield, the Grand Jury adjourned for the weekend after two days of testimony. Jury Foreman Carl Melcher, of McFarland, indicated that the jury had not concluded its work. District Attorney Dorsey, who had conducted the examination of witnesses, declined to make any statement. It was rumored that various of the subpoenaed witnesses had not shown up, among them Eli Andrews, George Bowman and a Fellows butcher who reportedly had received a threatening letter and was understood to have left town.

In Los Angeles, it was reported that William S. Coburn, an attorney said to be Grand Goblin of the Ku Klux Klan, had left town, reportedly to visit Taft after receiving a hurry-up call requesting his presence there.

On that same Friday night in Taft, a photographer for the *Los Angeles Examiner* in response to a message stating he would be allowed to photograph Klansmen waited in his car near the Standard Service Station on Center Street. At 7:30 P.M., a large car with curtains up drove alongside and an occupant advised the photographer to follow. The short drive led to the grounds of Conley Grammar School in South Taft.

At the school, Mrs. H. E. Tupper, who with her husband acted as caretaker, answered a knock at the door of the couple's basement quarters. The man at the door told her he wished to take some pictures. When Mrs. Tupper started to go outside, the man attempted to restrain her. Mr. Tupper, hearing the disturbance, came to the door and the couple went outside. There they saw hooded, white-gowned men standing by five automobiles. When Tupper started toward the cars, a spotlight was shone in his face. The Tuppers were forcibly turned around and told to go back into the school.

The picture posed by Klansmen showed one of the group in a black hood kneeling in the center of a ring of ten other Klansmen armed with rifles and pistols. The photographer said it was meant as a demonstration of how Klansmen surrounded a bootlegger who came before them.

On Monday, March 13, Private Detective Pyles with an assistant, Bert Tibbet, and two newsmen found the place on Midway Oil Company property near Maricopa where Druggist Bowman had been beaten. The party found two sections of one-inch rope, each about three feet long and knotted at the end, as well as a bag of feathers and an empty five-gallon square can that had contained tar. It was the first material evidence to be found.

On that same night, five federal prohibition agents under the direction of F. A. Hazeltine raided alleged bootleggers on the West Side and arrested nine men, including a Taft druggist.

Supervisor Stanley Abel accompanied the men on the raid and later said, "The supervisors of Kern County are tired of flagrant lawlessness in this district and the disregard local officials have for the violation of prohibition law." He used the raid as an occasion to make his first public comment on alleged Ku Klux Klan activity in the area. Stating that the raid had nothing to do with operations of the Klan, Abel said, "I have heard that armed gangs have been active here under the guise of the Ku Klux Klan, and no doubt local citizens aroused to action by the inaction of local officers have overstepped themselves to attain their own ends. However, I believe that Mr. Dorsey is now investigating these cases and it will be the duty of the supervisors to see that the law takes its course."

The Garner Adult Bible Class of Taft's Methodist Church passed a resolution highly commending the supervisor for his vigorous enforcement of the liquor laws, and the Reverend Van Dyke Todd of Fellows also praised Abel. Calling for strict enforcement of existing laws to remedy "conditions" in the oil fields, the Reverend Todd said, "I believe in the enforcement of the law through ordinary channels, and don't want to see people go outside the law until every other means is exhausted. However, I can see where the time might come when patience ceases to be a virtue, and I believe that the local Ku Klux Klan was organized to aid, not hinder just administration of the law."

While authorities continued the apparently fruitless search for suspects in recent beatings, the Klan seemed to undergo a curious metamorphosis on the West Side.

An unidentified man inserted a classified ad in the *Driller* which said:

"Would like to meet a representative of the K.K.K. This is straight stuff as the writer has a duty for them to perform. Address Box 100, *Driller* office, making appointment."

The man declared the Klan was doing good work in Taft and said he thought they could help him administer justice which could not be obtained in the ordinary way. The newspaper made no mention of response, if any, to the ad.

More than a dozen hooded men appeared late one night at the home of a South Taft man whose wife had died recently, leaving the widower with six small children. The night riders left two boxes of groceries and a note, with directions that the note be taken to the *Driller* for publication. The note said:

"The Ku Klux Klan extends to you their sympathy, and ask that you accept this little bit to tide you on until better times arrive.

"This will also advise you that whatever bills you owe for medical service have been paid by the Ku Klux Klan, so you may consider yourself free from this obligation. Should you find yourself in any difficulty of any nature you have only to tell one of your friends and we will know.

"You are cautioned not to part with any of your children without advice.

"You have our best wishes, and we will see that you have help when needed.

"Taft Klan No. 3, Ku Klux Klan, Realm of California."

A local stock company, Frank & King's Comedians, announced plans to present a play titled *Ku Klux Klan*. Described as a four-act drama, the play was to be given a four-day run at the Big Barn on Center Street just below First. Manager Jack Latham of the stock company said two men from the Ku Klux Klan had asked that they be allowed to censor the production at a rehearsal and he would permit them to do so.

The *Driller*, describing the play as one with a theme familiar to all who had read the well-known novel *The Clansman*, said it was a story of the South during the reconstruction period "when the newly freed slave of the South, by the power of the franchise given him by the North, filled all the public offices of the state, county and town with the riffraff of crooked politicians later being classed as 'carpet baggers.'"

"The plot of the piece," the *Driller* continued, "makes use of two heavies each of a widely different character but both made so despicable by the author and the art of Frank & King's company of talented artists that no regrets are apparent among the audience when at the close of the play and after the two villains have been dragged off the stage the two shots a few seconds apart announce that two more good Indians have been made."

At the premiere performance, the newspaper reported, as the drama was drawing to a close and while a tense scene was in progress between the leading lady and the villain, Toby Leland, well known in local stock, made his appearance. Another actor remarked, "If you don't look out, the Ku Klux Klan will get you." Leland replied, "I don't care," with what the *Driller* described as "carelessness bordering on recklessness."

The newspaper continued, "Just then thundering footsteps were heard in the lobby. Closer and closer came the tread of feet. The play came to a dead stop, and actors and audience alike froze.

"'That's the man,' said a hoarse voice. 'That's the man we want. Let's get him.'

"Terror filled the air. Under the seats went the larger portion of the audience and it seemed for a few awful moments that a panic could not be averted.

"'Get him, men,' said the leader, pointing a stern finger straight at the shrinking form of Leland as he cowered into a corner of the stage in an endeavor to escape the notice of the grim-visaged, white-clad men.

"Down the aisle strode the score of white, silent figures. Onto the stage they leaped and grabbing the unresisting Leland made their way toward the exit.

"One man braver than the rest managed to stop the leader. 'Where are you taking him?' he asked. 'And who are you?' he added. 'We're the Elks drill team and we want this man down at the Redmen Hall where we are holding a big meeting so that we can initiate him.' They left the building, Toby in tow."

The *Driller* reported that the Big Barn was filled for performances of *Ku Klux Klan*.

Another theatrical troupe made its debut in Taft with the opening of the Southside Theater. The troupe was Kelly's Komedians, quickly dubbed Kelly's Klever Komedians. The first play, *Her Unborn Child*, was a three-act melodrama staged with such realism that when the plot called for the heroine to fry liver, a functioning gas stove was provided and she actually fried the liver on stage.

On Tuesday night, April 4, almost two months after the warning handbills appeared, authorities seemed no nearer a solution to beatings on the West Side. No arrests had been made, nor had anyone come forward to name suspected assailants.

George Pettye, a bachelor who had formerly managed an automobile agency and now owned a small eating house on Center Street, was in his home washing dishes when two large men wearing white robes and hoods entered with drawn pistols. Pettye tried to flee through the back door, but was seized by men waiting there. The men placed Pettye in a car and headed toward Maricopa with another car following. About a mile out of town, the party stopped. They stripped Pettye and beat him with ropes. Afterward, the leader told him, "This is just a taste of what is to follow if you do not mend your ways." The leader told Pettye to remain in town, but said that a certain girl was to leave town within twenty-four hours. The hooded men drove back to Taft, leaving Pettye to walk. Friends said Pettye had received two warnings from the Klan within the past three weeks but had paid no attention.

Six nights later in Fellows, Lee Ellison, a local resident, was walking in front of the home of Gus Schoenfeld, manager of the Fellows drugstore, when a white-hooded man stepped out and in hushed tones advised him to move along, at the same time shoving a rifle barrel toward him. Ellison walked about a half block and doubled back. He recounted, "Just as I neared Mr. Schoenfeld's home on my way back, I saw perhaps a dozen white-clad figures skulking in the yard. I heard one of the men at the front door say, 'We want you.' Just then the lights in the house went out and Schoenfeld answered, 'Come in and get me.' There was a hurried consultation and a quick retreat to the cars, and the six big machines whisked away down the street, leaving Schoenfeld behind."

The *Driller* reported, "According to the story today, Schoenfeld saw the figures surrounding his home, secured his Winchester pump shotgun, turned out the lights and waited for the first uninvited visitor to enter. So far as could be learned, he has been given no warning of any nature." Schoenfeld, who was widely known to keep a loaded .45 calibre pistol by the cash register at his drugstore, reportedly let it be known that he would be waiting if his uninvited visitors chose to return.

On the same night, a band of hooded men believed to be the same group that had been at the Schoenfeld home rapped on the

front door of the Fellows home of Walter Hyat. Hyat came to the
screen door in night clothes. When he saw the men's masks, he ran
toward the rear of the building. The hooded men tore the screen
door from its hinges, entered the house and seized Hyat. They
placed him in a car and drove about two miles out of town. They
beat Hyat with knotted ropes, leaving him to walk back barefooted.
Hyat, a married man with two small children, reputedly had been
blacklisted in the oil fields following the recent strike and was said
to have been engaged recently in selling auto accessories. He reportedly had not received any warning.

In the wake of renewed violence, a Maricopa man named E. D.
Hoover addressed an open letter to the Klan through the columns of
the *Driller* asking for an audience with the Klan to discuss a warning
received by an unnamed young man of his acquaintance. Hoover
said the young man had been unjustly judged.

The Klan replied in the following day's edition:

"Regarding your letter appearing in today's *Midway Driller*, beg
to advise that we have no knowledge of any letter being sent to your
friend. Therefore we cannot possibly have anything to discuss with
you. However, we will greatly appreciate your co-operation in ap-
prehending the sender of the purported letter.

"Our motto is 'equal justice to all,' and we insist on this for ourselves
and will accept no less from our enemies.

"If you will send your information, as you will know how to do so,
we will give it the attention that it calls for.

"Yours very truly, Taft Klan No. 3, Knights of the Ku Klux Klan,
Realm of California."

On Monday night, April 17, Private Detective Pyles, who had
offered to expose the Klan, was informed that a meeting was in
progress in a canyon near Maricopa that might involve I.W.W. matters he should investigate. Pyles had represented oil operators in the
oilworkers' strike during which I.W.W. involvement had been
alleged.

Pyles drove to the canyon with three of his men. After the men
had taken cover, Pyles approached a man who appeared to be a
lookout. The man challenged him. Pyles later gave this account of
what followed:

"I quickly whipped out a revolver and told him to stick up his
hands and keep still ... The rest of the gang, attracted by the guard's
challenge probably, closed in on me. Gun play was useless, so I

stowed my gun. Soon then I was encircled by 50 or 75 masked men, all heavily armed and threatening. They conducted me into the hollow, a short fellow, probably the leader, taking charge. All this time the masked men kept prodding me with their guns, threatening me with death.

"'Did you take any of our robes?' the leader asked. Then I knew what organization was meeting, although I did not take any of their robes. I answered 'no,' and this infuriated the short man. My hands, in the meantime, had been tied behind my back with wire, which cut into my flesh. He asked me several other questions, which I will not repeat because it will not help the situation, and then measuring off struck me on the point of the jaw with his fist. Struggling to my feet, I called him a coward, and he hit me again. I fell in a heap, helpless, and painfully bound. But this cruelty seemed to have no effect on the rest of the party.

"Weak from the two blows, I was conducted to a car and driven to a spot which I afterward learned was about 12 miles from Maricopa in the direction of the San Emidio ranch. I was cursed, struck and shoved during the trip, more perhaps than I would have been had I not returned their curses with equal heat about their cowardice.

"Three men were in the car, including a big, burly fellow who seemed to be the chairman. I think that this was a committee sent to beat me, as I afterward learned that the meeting continued until midnight, the time I got back to Maricopa.

"It was about 11 o'clock when we reached the spot where the beating took place. I was told to get out of the machine and walked around in behind. I did so, not knowing what was going to happen. The big fellow and his two companions withdrew out of hearing and went through a ceremony, which I believe is a sort of ritual or prayer given before a beating.

"Then they came after me, the big fellow roughly shoving me from the car. He then administered the beating, with a heavy rope. Time after time he lashed me, until I could no longer keep my feet.

"'Will you leave the county in five days?' he said at length, ceasing the whipping. Between my teeth, I answered, 'No.' 'Well, take this and that and that, you dirty _____.' I collapsed, and when I recovered my tormentors had departed.

"Crawling, stumbling, blinded, my head buzzing, I struggled along until I came to a road. Here I was picked up and brought to Maricopa, where I was treated. One of my men was also caught, but for some reason, he was not beaten. I was unable to work my blindfold loose and for this reason could not identify my assailant, but I'm not done yet. Kern County must be ridded of this menace to its peace and dignity."

On the following day, a businessman found a penciled note under his door. The note read:

"An appetizer for breakfast:

"J. N. Pyles, our famous detective, tried to capture single handed 100 K.K.K.'s while they were having a meeting but he met with hard luck and was severely beaten. He spent an hour at a hospital getting repaired and then went to a hotel.

"Thad Cheney, Pyle's brother and another man were with him. The 4th party was caught and the brother and Cheney are still looking for John."

On the Saturday following the assault, Pyles returned to Taft. Limping badly, he carried a sawed-off shotgun. He said he would not leave the fields by May 1 as ordered to do on pain of death, and intended to turn over to District Attorney Dorsey information that would lead to sensational developments when the Grand Jury convened in Bakersfield on the following Monday.

On that same Saturday night in the Southern California city of Inglewood, Frank Woerner, the city's night marshal, responded to a call that the home of the Eiduayen brothers, who ran a licensed winery, was being threatened by an armed and hooded mob.

Woerner later testified, "I rode down there on the motorcycle with Clyde VanNatta. We were stopped by some men who told us to throw up our hands. I said: 'I am an officer; throw up your hands.' One of them with a gun in his hand started for VanNatta. I wanted to protect the boy's life, so I shot this fellow. Then a couple of them started for me. I wanted to protect my own life, so I shot them.

"Everybody was shooting at me, and somebody hollered, 'He's killed one of our boys. Get him.' I kept on fighting until I ran out of ammunition. Then I beat it. I thought they were hold-ups, because they wore masks."

When masks were removed, the dead man was identified as a constable. One of the injured men was the constable's deputized son. The other was a deputy sheriff. All were members of the Ku Klux Klan.

Four days later, Chief Deputy District Attorney W. C. Doran of Los Angeles County and Undersheriff Eugene Biscailuz, armed with a search warrant, led a raid on the office of Attorney William S. Coburn, Grand Goblin of the Klan, and seized all visible Klan property, including membership lists with the names of approximately 1,090 Klansmen.

Los Angeles County District Attorney Thomas Lee Woolwine issued a statement: "It would seem to me that no right thinking American could find the slightest excuse for the existence in this country of an organization such as the Ku Klux Klan. The name itself has a sinister and well known meaning. It breathes the spirit of the night rider and of cowardly assassinations and terrorism. It perpetrates its dastardly outrages in the dark hours of the night, caparisoned in robes of terror and burglars' masks. Its membership is generally kept secret, there being no insignia regularly worn indicating membership in any secret order as in organizations that are lawful and above board and walk in the daylight."

On the day following the raid, District Attorney Dorsey closeted himself with District Attorney Woolwine and Deputy District Attorney Doran in Los Angeles. A staff of photographers was placed at Dorsey's disposal so that letters and other material bearing on the situation in Kern County could be copied and copies certified for use.

Dorsey said, "This is what I have been waiting for. For weeks I have been holding a great deal of inanimate evidence—tar buckets, masks, weapons, and the like—but I have never been able to secure enough positive evidence to cause the arrest of men I have suspected.

"Now through this evidence, it will be possible for me to lay positive proof before the Grand Jury. Some of the names revealed are those of prominent, popular residents."

On Monday, May 1, the *Driller* reported that startling evidence had come to light. The newspaper said that according to District Attorney Gearhart of Fresno, who had visited Woolwine in Los Angeles, a "boastful" letter written by a Klansman in Taft to the Grand Goblin said that Dr. Dwight R. Mason, of Taft, had been taken out to answer to certain accusations, which later were found to be untrue. A knife was thrust into Dr. Mason's back, the Klansman wrote, while other Klansmen stood around and attempted to force a confession from Mason. Every time Mason denied the charge, the knife was twisted.

At that night's session of Taft's Board of Trustees, Porter Munsey submitted his resignation as city marshal. The board named Roscoe C. Steele to succeed Munsey. Steele, a newcomer to Taft, was said to have long experience in official capacities, having recently served as a member of the Santa Barbara police department. The board empowered him to swear in a new slate of deputies.

While rumors persisted that the names of those belonging to the Klan would soon be released, hearings continued before the Grand Jury in Bakersfield. When E. A. Abbott, president of the Kern County Building Trades Council and an admitted Klan member, refused to answer questions, citing the Klan's secrecy oath, Judge Harvey ordered him locked up.

Judge Harvey said, "The law of the state of California will not strike its colors to any man, or any organization . . . No individual and no order can come into my court and say because of an oath they have taken as a member of a secret organization they will not answer and get away scot free and unscathed . . . Anyone who cares to defy the spirit or the letter of the law in my court will meet the same fate that I have meted out to E. A. Abbott."

After several hours in county jail, Abbott, a plumber and Sunday School teacher, decided to answer questions.

In Taft, the Southside Theater advertised that its next attraction would be *The Days of '49*. The stock company which formerly had been called Kelly's Klever Komedians was identified as Kelly's Players.

In Bakersfield, Dr. Mason arrived from Los Angeles under protection of three armed guards to make a statement to District Attorney Dorsey regarding the treatment he had been given on the night of October 27, 1921. In the statement, Mason said he knew the men who tortured him and they were members of the Ku Klux Klan. He said he had been dragged from his bed in the middle of the night, taken to the ball park and hanged by a rope thrown over a rafter. He said he was given three distinct whippings while partially conscious.

Describing the hanging, Dr. Mason said, "I thought my end had come. Then I lost consciousness. When I came to, I was lying upon the ground. But I was not left there long. The whippings followed and I thought the pain of the knotted ropes and wires would kill me.

"My back and legs are covered with scars which will remain with me all my life.

"Following the whippings I was allowed to crawl away. However, I was warned of 'certain death' if I told the authorities a word about the treatment which had just been given me.

"Fearing for my life, I had since kept silence."

Dr. Mason told the District Attorney that the attack on him was the sequel of a suit he had filed in Superior Court of Los Angeles County in which a woman had been made the defendant. The woman, he said, was present at the hanging, encouraging the mob.

On Monday, May 8, the *Driller* published the names of 132 men identified as Ku Klux Klan members on a list released by the District Attorney.

The list included Stanley Abel, chairman of the Board of Supervisors, Mayor H. C. McClain; City Trustees J. M. Higgins and Herbert V. Hearle; Justice of the Peace George M. Cook; City Clerk C. Z. Irvine; Porter Munsey, former city marshal; R. M. Padrick, assistant fire chief; and others ranging from professional men to oil-workers and including the Reverend Van Dyke Todd.

In the same issue, the *Driller* printed a denial by City Trustee Haskell M. Greene that he was a Klan member. Greene had been identified as a member by the *Bakersfield Californian*.

On the following day, Mayor McClain, a barber, announced his resignation from the Klan, issuing the following statement:

"I belonged to the order for the principles the obligations stood for, and I have lived up to that obligation as near as I knew how, and if anyone in the organization has committed a crime they have gone far from their obligations."

The *Driller* published statements from various of those named declaring they had joined the Ku Klux Klan more than a year before at a time when little was known about the Klan and subsequently had not taken part in its activities. Some said they had paid a $10 fee to a Klan organizer upon representation that the Klan was founded upon principles for the betterment of America, but had never taken the initiation, paid dues or attended meetings.

Among statements was one by City Trustee Hearle that he had attended two meetings about a year ago but had paid no dues and therefore considered "that I had nothing to do with the order."

Supervisor Abel issued a statement which said:

"Yes, I belong to the Knights of the Ku Klux Klan, and I am proud to be associated with many of the best citizens of Taft and vicinity in the good work they are doing. I know nothing of the activities of the

Klan in other localities, but the Klan at Taft is certainly deserving of praise for the good work it has done in ridding the community of the class of scoundrels who were selling bootleg whiskey and doped candy to high school boys, and others, attempting to debauch the young womanhood of the community.

"We have tried to cooperate with the proper authorities, but they refused to do any effective work.

"Grand juries are powerless unless they have cooperation from the district attorney's office. The bootleggers in McKittrick openly boasted that the sheriff's office would tip them off in case of a raid.

"The raid was made by federal officers, assisted by citizens of the West Side and was a complete success. Seven places were selling bootleg whiskey to any who would buy.

"The report of the grand jury last year refers to the fact that they did not receive any cooperation from the district attorney.

"Good people cannot and will not stand idly by after repeated efforts to get the law enforced, and see the boys and girls of the community debauched by lawless aliens who curse the constitution and defy our laws.

"I know nothing of any lawless act committed by the Klan or its members. Those who accuse the Klan or its members of lawlessness should place their information in the hands of the proper officials.

"I make no apology for the Klan. It needs none."

On Tuesday night, May 9, scarcely twenty-four hours after publication of the Klan's membership list in Taft, an estimated 3,000 persons held a mass meeting in the Sunshine Theater at which the following resolution was offered:

"WHEREAS: Recent developments baring the activities of the Ku Klux Klan have shown that a number of state, county, city and school officers have been involved in said activities, therefore, be it

"RESOLVED: That it is the sense of this meeting that such activities are in absolute violation of the oath of office taken by said officials and absolutely un-American, and be it further

"RESOLVED: That all such officers, whether state, county, city or school district, be asked to resign from their respective offices."

When a rising vote was called, the whole group seemed to stand. On the call for those opposed, only two stood, both women. Booing and hissing greeted them, and they quickly sat down.

Two days later, District Attorney Dorsey revealed the names of fifty-six additional Klan members in Kern County. The list included the names of three Kern County Deputy Sheriffs, who were quickly ousted from service, as well as the name of a W. N. Thompson. The latter brought an emphatic response from W. N. Thompson, Standard Oil Company's production superintendent at Elk Hills, declaring that the man named was a Thompson other than himself. "I consider myself too good an American citizen," Thompson stated, "ever to be connected in any way with any organization such as the Ku Klux Klan that does not try to right what it considers wrong in the open."

In Bakersfield, the Grand Jury indicted John H. Vitelle, Taft contractor, on three counts, including one charging him with use of a gun upon the person of Dr. Mason, another with taking part in the hanging of Dr. Mason and a third in taking an active part in beating Dr. Mason with a knotted rope. Vitelle was identified as a charter member of the Taft Klan and the Klan's Exalted Cyclops, or president, at the time of the assault on Dr. Mason.

When Klan membership lists were published in other communities outside Kern County, a shock wave struck Taft. Roscoe C. Steele, who had been named city marshal early in May to succeed Porter Munsey, was a member of the Santa Barbara Klan. The nine deputies Steele had been empowered to appoint were also identified as Klansmen.

On Friday, May 19, a delegation of approximately 150 persons drawn from every walk of life on the West Side set out for Bakersfield. The caravan began to arrive about 1:30 in the afternoon at the Kern County courthouse. Parking along Chester Avenue, the group formed a body and moved into the courthouse, asking to see District Attorney Dorsey.

Dorsey met the group in the main floor lobby of the building. The gathering formed a circle around the District Attorney and Carl Melcher, foreman of the Grand Jury, while overhead, galleries of spectators lined the huge circle opening through the second and third floor landings.

Dr. Ernest Ballagh, Taft dentist, spoke first, saying, "We have come to tell you that the people of the West Side do not want the Ku Klux Klan, and that you have our undying support in your efforts to stamp out this organization that is a menace to our community."

Jack Hamilton, official spokesman for the delegation and former school superintendent in Taft, delivered a brief message. He said, "Whether we are your political enemies or your political friends, or whether we are your personal enemies or your personal friends, we stand united on this fight. We want a complete investigation and a thorough cleaning out of this secret organization.

"This delegation represents the citizenry of the West Side. Look the crowd over," he said, waving his hand around the human circle that surrounded him, "and you can judge whether we are bootleggers or any such gentry as that. Some are businessmen, professional men and working men, upbuilders of the community, making this declaration against the Ku Klux Klan and in so doing we have the backing of practically all of the people of the West Side."

District Attorney Dorsey expressed his appreciation for the demonstration, calling it "a great manifestation of Americanism."

At a special meeting of Taft's Board of Trustees on Thursday night, May 25, J. M. Higgins resigned as trustee. At the same meeting, City Marshal Steele and Assistant Fire Chief Padrick submitted resignations. Former Constable Arthur Turner was appointed to fill the marshal's post.

Early in June, a citizens' committee initiated recall proceedings against Supervisor Abel, Justice of the Peace Cook, Mayor McClain and City Clerk Irvine.

On June 26, 1922, the trial of John H. Vitelle began in Bakersfield. On June 30, Vitelle was found guilty on one count of assault, namely, flogging with a heavy rope with intent to inflict bodily injury to the person of Dr. Mason. Five days later, he was sent to San Quentin prison to begin serving a one to ten-year sentence. He was the only member of the Taft Klan to be tried for his role in incidents attributed to the Klan on the West Side.

On the day after Vitelle went to prison, the Taft Klan addressed an open letter to the public. The letter said:

"This is to certify that Taft Klan No. 3 was disbanded June 16, 1922. We wish to take this opportunity of saying that we repudiate any act of lawlessness committed in the name of the Klan and that we will do our utmost to bring to justice anyone committing such acts.

"It is further certified that all robes were destroyed on this date and that all equipment was delivered by me in Los Angeles."

The letter was signed by Frank M. Page. Page, an employee of Midway Oil Company, described himself as former Exalted Cyclops.

At the recall election on August 29, Justice of the Peace Cook, Mayor McClain and City Clerk Irvine were turned out. The vote was as follows:

Cook, for recall, 1,277, against recall, 953; McClain, for recall, 584, against recall, 363; Irvine, for recall, 590, against recall, 409.

Supervisor Abel, who two years before had been elected to office with a majority of more than 1,200 votes, escaped recall by 103 votes. The vote for recall was 1,799; against, 1,902.

Disaster lay in wait, but there was no hint of the impending holocaust when this picture was taken on the floor of Milham Exploration Company's Kern No. 1. (Photo by Dalton Gautreaux)

On the afternoon of election day, Tuesday, November 2, 1926, Kern No. 1 blew out, catching fire almost immediately. (Photo by Dalton Gautreaux)

A Most Prankish Well

There had been signs of trouble at the well, but nothing so threatening as to suggest what was about to happen. If there had been, Sam James, for one, might have had second thoughts about going to work. James was the derrickman. To perform his duties, he had to stand on a small platform high up in the steel derrick, leaning into the bellybuster to rack drill pipe being pulled from the hole. It was a position that shortly would put him almost directly over the mouth of the 15½-inch diameter casing which had been cemented in the ground to protect the hole. The casing was about to be transformed, in effect, into the wide muzzle of a cannon with greater fire power than the dreaded railroad guns that had spread death and destruction on the Western Front in the World War.

It was election day, Tuesday, November 2, 1926, a pleasant fall day which the weatherman had correctly forecast as "fair and mild, with light variable breezes."

The election pitted Lieutenant Governor C. C. Young, a Republican, against Justus S. Wardell, a Democrat, for the governorship of California. Republican registrants outnumbered Democrats by a comfortable margin. Of nearly two million Californians registered to vote, 1.3 million were registered as Republicans, 410,000 as Democrats. There was little doubt who would win and, in fact, the election ran true to form with Republicans sweeping the state, a compliment to Calvin Coolidge, the Republican who occupied the White House.

Of more interest than politics was the football game to be played the following Saturday afternoon in Bakersfield, matching Taft Union High School's Wildcats against Bakersfield High School's Drillers. It was the big game between traditional rivals. The Drillers, state champions, were heavy favorites, but the Wildcats' Coach Crip Toomey was hinting at an upset. It was rumored that Toomey was shifting his backfield of Cameron, Johnson, Nesbit and Clark and might have some surprises for the Drillers.

At Milham Exploration Company's Kern No. 1 near the farm community of Buttonwillow, seventeen miles north of Taft, Sam

James and others in the five-man crew were making a trip. They had
drilled to 3,323 feet in what was programmed to be a probe to 7,000
feet and were coming out of the hole. There had been gas shows in
the well, including one not much more than a week before while
the crew was pulling a core from 2,669 feet. The well had kicked,
but the gas blowout had lasted only two or three minutes before the
crew brought the well under control. Since then they had kept
heavy mud in the hole. While there had been continuing shows,
there had been no repeat of the short-lived blowout. The drillers
were experienced, and the big steam rig they were working with
was more than adequate for the drilling assignment. The powerful
rig was something of a showpiece.

At 4:30 P.M., while the trip was in progress, Kern No. 1 blew out.
The first warning came when gas pressure began to push drill pipe
out of the hole. On the rig floor, Driller B. C. Brown and the three
roughnecks ran for safety. Up the derrick, Sam James endured the
unnerving experience of watching some 400 feet of heavy drill pipe
fly past him in a shower of blue mud. Writhing like a snake, the pipe
ripped out the top of the steel derrick and fell in a tangled mass like
so much spaghetti a hundred feet away. Miraculously, James escaped death. He dropped like a plummet down the safety line and
joined others running as hard as they could across the field in a
shower of mud and debris.

When the men looked back, flying sand had struck a spark in the
casing. Gas became a livid tower of flame shooting high into the sky
above the top of the 125-foot steel derrick. Within minutes, heat
exploded an adjacent oil tank and a water tank. The earth shook,
and the air was filled with the sobbing roar of escaping gas.

In the small community of Buttonwillow, eight miles south of the
well, ground trembled as if in an earthquake, except that the shaking
did not stop. When people rushed from their houses, they heard a
distant roar. Looking toward the sound, they saw a towering column
of smoke climbing into the sky.

In Los Angeles, S. J. Hardison, manager of Milham Exploration
Company, received a telephone call advising of the blowout. Without delay, he climbed into his car to begin the trip to the drill site.
The journey into the San Joaquin Valley over the Ridge Route with
its 642 narrow radius curves and seven per cent grades allowed
ample time to speculate on what Milham had gotten into. Had the

Boilers powered the rig that put down Milham's Buttonwillow wildcat. It was to be the last job for the rig and boilers. (Photo by Dalton Gautreaux)

company discovered a major oil field? Or was the blowout just another gas blow that quickly would run its course?

Milham had come to the Buttonwillow prospect area early in 1926, initiating geological work under the direction of L. S. Chambers, field geologist. The company enjoyed the reputation of being one of the best financed companies operating in California. One of the principals in the firm was John Hayes Hammond, who had formerly been with Cecil J. Rhodes in South Africa and was a mining engineer and financier of world-wide fame. Hammond resided in Washington, D.C., but was well known in the San Joaquin Valley. In geological surveys, he had visited Randsburg and other mining districts. He was said to have assisted in financing Mt. Whitney Power Company in Tulare County and, like Herbert Hoover, had been largely interested in California through early mining associations.

Other companies had preceded Milham in the Buttonwillow area. In 1921, the area had received a boost from publication of an article by Roy Ferguson in the September issue of the Division of Oil & Gas' *Summary of Operations*. Ferguson, an experienced geologist on the staff of the state agency, had reviewed evidence of anticlinal structure, summarized results of early drilling and concluded that indications were "encouraging" for a discovery.

In the same year the article appeared, three wildcatters began work. Petroleum Midway Company, Limited, drilled on the Laird property, abandoning the hole following a blowout that froze drill pipe at 4,034 feet. National Exploration Company drilled an unsuccessful wildcat. While the latter hole was under way, Petroleum Midway drilled Bush No. 1 as a follow-up to the Laird well. The Bush wildcat blew out at 3,500 feet. Fragments of lignite blown out of the well were similar to lignite found in the oil zone at Elk Hills, fourteen miles to the south. The well was tested, but it failed to produce and subsequently was abandoned at total depth of 4,435 feet. Shell Oil Company initiated a drilling campaign, eventually putting down fifteen shallow holes, most no deeper than 1,000 feet, and one deeper test that went to 4,628 feet. There were shows, but none of the wells found commercial production, and the company quit the campaign.

Although the drilling flurry in the early 1920s did not turn up a successful completion, the search was not allowed to die, partly because of events on the nearby Semitropic Ridge, where Main Oil Company in March 1923 spudded in to drill its No. 1 wildcat, seven miles northeast of where Milham later would drill.

Gas shows started at about 3,200 feet in the Main wildcat. At 3,970 feet, gas blew fluid over the crown block. There were strong gas shows to 4,300 feet, and practically continuous shows from that point on to 5,000 feet. The company went to total depth of 5,305 feet and brought in the well under control late in April 1925. Pressure was so great that fittings controlling the well quickly wore away, and gas accompanied by water blew wild. Blue shale spewed from the well.

The flow of gas, water and shale stopped after twelve days. Numerous Scalez petrolia were found on the ground in pieces of shale. Scalez petrolia, a small flat fresh-water calcareous shell found in the upper portion of the Etchegoin formation, earlier had been identified as an important marker in the Elk Hills and Buena Vista Hills fields, where it generally overlaid oil sands.

Main Oil Company rebuilt the derrick over the well and cleaned out the hole to 4,300 feet. Several blowouts occurred during the work. When efforts were made to produce the well, there was only a small amount of gas. The well was suspended in May 1926. The company moved some 700 feet to the southwest to drill a follow-up. The well went to 5,272 feet. There were no shows.

When Milham tackled the Buttonwillow prospect, the company found itself working an area that had not shared in the prosperity from oil enjoyed by other West Side towns. The unincorporated community of Buttonwillow, which took its name from a solitary buttonwillow tree under which Yokut Indians were said to have gathered for trading sessions, seemed not much more than a wide spot in the highway with a filling station or two, a post office, an eating establishment and a reputation as a place where a man might buy a bottle of bootleg whiskey.

While West Side fields contributed the bulk of the oil that kept the state of California number one in the ranks of oil producing states, the Buttonwillow sector remained without production. Each week the American Petroleum Institute in New York released production figures that underscored the have-not status of Buttonwillow in an otherwise bright picture. On the very day Kern No. 1 blew out, the API was quoted in newspapers with a release showing California well above any other state with a production of 628,300 barrels per day, which put the state comfortably ahead of its nearest rival, Oklahoma, with an output averaging 539,300 barrels daily, and even farther in front of the next nearest rival, Texas, which was credited with 460,350 barrels a day.

Milham, after careful geologic study, opened its Buttonwillow campaign with a pair of shallow holes drilled primarily to gain structural information. On September 26, 1926, the company spudded in on Sec. 8, 28S-23E, to drill Kern No. 1 on a topographic high. Programmed for a depth of 7,000 feet, the wildcat shaped up as the deepest to be drilled in the Buttonwillow area.

After the blowout, flames shot over the top of the derrick in a fire which observers would later describe as one of the biggest and most spectacular gas fires in the history of California.

The fire raged uncontrolled, blasting from the gaping mouth of the well, leaping in a wide column an estimated 600 feet into the air. The steel derrick was torn to pieces.

During the first hour, huge chunks of blue shale and mud burning hot from the blaze showered the vicinity for a radius of half a mile. A massive plume of gray and black smoke drifted in the sky, broadcasting news of the fire throughout Kern County.

When night came, leaping flames could be seen for a distance of more than twenty miles. Clouds above were bathed in crimson. The spectacle attracted thousands of spectators, who clogged dirt roads with their automobiles, eager to get a close view of the giant inferno that offered a frightening glimpse of nature running wild.

One of the visitors to the scene of the blowout was George Suman of Associated Oil Company, who estimated the well was making at least 20 million cubic feet per day of gas. "One could not get within several hundred feet of the burning gas well," Suman said, "but even at that distance the noise was so great from the escaping gas that it was impossible to hear the horn of an automobile. The crumpled derrick and the havoc of the terrific force could easily be seen from a distance."

Frederick E. Hoar, who visited the site at night, said a continual shower of blue shale was being shot from the hole, some pieces being as large as a football. Hoar said the ground for a distance of 500 to 600 feet from the hole was subjected to constant bombardment and the roar was terrific.

Hundreds of oilworkers from Taft, Elk Hills, McKittrick and the Belridge oil field rushed to the scene, eager to be of assistance if a fight were to be mounted to control the fire.

Among visitors were scores of geologists, representing practically every oil company operating in Kern County, who were intent on examining sand and shale spewed from the well, searching for the fossils that might give valuable clues to the possibility that the region was oil territory.

In the town of Wasco, fourteen miles to the northeast, the night was lurid in the glare of distant flames, and one could read a newspaper in the intense yellow light.

If Sam James or any of the others in the drilling crew cared to read a newspaper that night, it seemed probable the part they would have found most interesting would have been the advertisements announcing new model cars. Chandler Sales & Service, a Cleveland, Ohio company, called its 1927 models "the most impressive six-cylinder motor cars the industry has yet produced" and said the standard coupe included "'one-shot' centralized lubrication, oil purifier, air cleaner, 4-wheel brakes, thermostat heat control, new type easy steering and self-adjusting spring shackles." A Touring Standard Six could be purchased for as little as $945. Hudson described its offering as "the finest Hudson ever built" and said features included "improved gasoline performance, new bodies in two-tone colors and the super-six motor." Price of the Hudson Coach was $1,095. Oakland Six advertised "rubber-silenced chassis, harmonic balancer and tilting beam headlights" in models priced from $1,025 to $1,295.

The advertisements had special meaning for men in the Milham drilling crew. When the well roared in, the accompanying blaze not only destroyed the rig but also burned to a cinder the three cars in which the men had driven to work.

In Bakersfield, thirty miles east of the burning well, the heights of the Kern Bluffs on the north edge of the city afforded a magnificent view of the spectacle. The great column of fire stood high in the air, and it could be seen billowing and spouting as the stream of gas shot upward.

From the bluffs, the fire appeared to be of a reddish tinge, as if oil or water might be a part of the fluid escaping from the hole. The column appeared wider than the flame of the giant gasser, Hay No. 7, which had burned for ten days at Elk Hills in 1919 and which also had been visible from Bakersfield. The Elk Hills gasser had seemed more incandescent and also had appeared to burn in a fixed position like a column of fire.

It seemed plain enough that the Milham wildcat was on a productive structure of some sort. The question was whether it was an oil structure. In searching for comparisons, most compared the flaming well with the Hay No. 7 which had burned seven years before at Elk Hills. Elk Hills had turned out to be a major oil field, certainly one of the largest in the state. Had Milham found another Elk Hills?

By midnight, the Kern No. 1 had cleared itself of clogging sand. Fire raged to its greatest intensity. Oil men estimated gas flow at 50 million cubic feet per day. Heat became so intense the well could not be approached within a radius of 300 yards.

Capping the gasser appeared impossible. The rig was a shambles. The well was cratering. The only hope appeared to be that the well might sand up long enough to allow the wreckage of the rig to be cleared away and, hopefully, a valve to be placed atop the 15½-inch casing, assuming that somewhere in the debris the casing could be found.

An end to the spectacle came at 5:00 A.M. on the following morning, 12½ hours after the well blew in. A cave-in choked the tremendous gas pressure, killing the flame. Hardly had ground above the well cooled than a geyser of sand rose hundreds of feet into the air and the gas roared forth with seemingly increased fury. Gas did not

Shattered steel and a roaring, churning maelstrom of sand, shale and gas wrote finis to Milham's Kern No. 1. (Photo by Dalton Gautreaux)

When the fury abated, there was only a deep crater and a twisted mass of steel to mark the site of Kern No. 1 near Buttonwillow. (Photo by Dalton Gautreaux)

immediately ignite. For miles around, the well made itself heard in a reverberating roar. For a distance of several miles, ground trembled as if in an earthquake. Motorists declared their cars were badly shaken on close approach, and one said the trembling of the ground stalled his engine.

At 9:49 A.M., the well ignited for the second time with an explosion that shook the countryside. In Buttonwillow, the shock knocked tools to the floor in the San Joaquin Light & Power Corporation's plant. S. P. Cokely, whose home was eight miles east of the well, said the explosion shook the earth violently beneath his house. Farmers in the Poso Creek district eight miles north of Wasco reported that earth tremors were felt and that the explosion could be heard there. With increased fury, the well shot a mammoth jet of flame nearly 1,000 feet skyward for forty minutes. At 10:30 A.M., the second outburst ceased abruptly.

Within half an hour a third eruption took place, the most violent since the well blew in. In Wasco, houses trembled, dishes rattled and pans crashed to kitchen floors. Around noon, the gas flow began to abate, giving rise to the hope that the well might sand up again.

On the following day, the *Bakersfield Californian* in its Thursday, November 4, edition reported:

"Tired of cutting capers and roaring to the countryside, the Milham Exploration gasser southwest of Wasco is drowsily blowing away today. For more than 12 hours it has spouted gas without catching on fire. The most spectacular thing it does is to toss fragments of shale into the air spasmodically.

"But the Kern No. 1, as the well is known, will go into history as one of the most prankish wells in Kern County.

"More than a week ago it came in with a gas pressure that was controlled by the drillers. It broke loose again, and was controlled again.

"The drilling crew came to look upon the casing mouth as the entrance to a 'bad' well. It couldn't be trusted.

"Then came the big blowout two days ago when the well spat out 400 feet of drilling equipment, tried to kill a driller in the tower, and wrecked the steel rig after catching fire. The fire was one of the biggest ever seen in the state. The heat from it melted the outfit's big boilers as if they'd been made of lead.

"It was a good joke, too, to quiet down and stop burning for half an hour or so and then burst loose with an explosion that shook the area for 20 miles around. That was yesterday morning, and Wasco trembled and rattled as if in the grip of an earthquake.

"Throughout yesterday afternoon the well had a grand time sanding up, catching fire, sanding up again and catching fire again. But shortly after nine o'clock last night it called it a day, sanding up and extinguishing its blazing topknot."

By Friday, three days after the blowout, the Kern No. 1 had become a mound builder, fast burying itself beneath debris. The mound appeared to be as high as the derrick floor and as wide as a city block. In the center, there was a crater estimated to be some 80 feet across and 70 feet deep.

Hundreds of visitors continued to flock to the site. Though gas was flowing at a diminished rate and without burning, it was still dangerous to approach any closer than 200 feet, making it impossible to obtain samples of the gas so that it might be studied for petroleum content. It was estimated that the cost to Milham would be $100,000, including loss of the hole and drilling equipment.

On Saturday, while gas continued to flow from the great mound, Taft's Wildcats met Bakersfield's Drillers on the football field at Bakersfield. The Drillers completely dominated the game, winning by a score of 82 to 0.

One week after the blowout, cave-ins had largely checked the flow of gas, though some still escaped. Milham Exploration began assembling materials for a second rig to drill another well. Others scrambled to get leases in the area. One mile southeast of the Milham gasser, Hoyt S. Gale leased 80 acres comprising the south half of the southeast quarter of Sec. 9, 28S-23E. Landowners Paul and Jesslyn Packard and Lila M. Reed retained one-eighth royalty in any production that might be found.

Speculation continued as to what the Milham well meant. Because of the shallow depth at which gas had been encountered, some thought prospects were good that an oil structure would be discovered at less than 5,000 feet.

Others believed that a giant gas field had been found. If so, it would be the first in California. Gas had been found, but always in conjunction with oil.

John P. Hight, Jr., president of Royal Petroleum Corporation, San Francisco, visited the West Side to give attention to the development of Royal's 45,000-acre leasehold in Cuyama Valley and, in passing, checked into the situation at Buttonwillow.

Hight said, "I am certain that there is oil to be found there, and that the gas wells of Milham and Main Oil Company were edge wells and not far from valuable petroleum deposits.

"Because the gas of the Milham well was termed a 'dry' gas does not lower my faith in the district. Any petroleum gas that is sufficiently filtered becomes 'dry' and I believe that the latest gasser blew a gas that had been highly filtered through structural formations."

Eight months later, those wondering what Milham had discovered at Buttonwillow had their answer. Milham completed Kern No. 1-A, a follow-up well, flowing dry gas at a rate of 5,242,000 cubic feet per day through a 28/64-inch choke from the interval at 2,361-2,648 feet. The company had found California's first commercial gas field.

When Mice Attacked

A heavy rain had turned the bed of Buena Vista Lake, which had been reclaimed for agriculture, into a desolate sea of mud on the dark night of November 24, 1926, when the advance column of an unlikely army moved out in one of the strangest assaults ever launched. The task force, composed of thousands of house mice, was the vanguard of an army that was to number more than 30 million, an army larger than any force put together by men. In the line of attack as mice scurried from muddy burrows was the town of Taft, seven miles southwest of the playa lake bed, a community of some 5,000 population that lately had claimed for itself the title, "The World's Biggest Little City."

The round-the-clock oil business that gave the city its reason for being was in full swing even as mice began moving out. Forty-seven drilling rigs were active in the Midway-Sunset field that surrounded the town, working to put down the wells that would enable operators to take out of the ground even more oil than the 94,000 barrels a day presently being produced. The amount of oil was enough to make the field second in the state only to the booming Long Beach field which was enjoying flush production of 97,000 barrels daily in the wake of Shell Oil Company's Signal Hill discovery in 1921.

For those of Taft's citizens not working in the oil fields that rainy Wednesday night, it was a fairly typical weekday night in the energetic little city, with choices for entertainment, or simply something to do, including a public dance, a card party, professional boxing, and movies at three theaters, not to mention fraternal, church and union meetings.

While mice began their march, unsuspecting Taftians danced to the music of the "Ruf Nex" band at Buchanan's Pavilion. For card players, there was Five Hundred at Odd Fellows Hall, sponsored by Taft Review No. 41 of the Woman's Benefit Association. A highlight of the evening was the raffling off of a cedar chest. At the Sunset Athletic Club at First and Center Streets, Angel de la Cruz, billed as the fighting Filipino, was winning an eight-round decision over

Frankie Novey in the headliner of a five-bout card that the *Daily
Midway Driller* in its next edition would describe as the best card
ever offered in Taft. The action-packed main event followed Promoter A. A. Crosbie's six-round semifinal in which Cracker Warren
and Jerry Carpenter fought to a draw, earning prolonged applause.

For those who liked entertainment on a screen, there was a choice
of three shows. At the Hippodrome, a theater party for the benefit
of the Elks charity fund was underway. The feature film was *Battling Butler*, starring Buster Keaton. There were shows at 7:00 and
9:00 P.M., and the price of admission was fifteen cents for youngsters and fifty cents for adults. At the Sunshine Theater, the evening's fare consisted of Richard Dix in *The Quarterback*, an action
picture which had been filmed with the technical assistance of
Fielding H. "Hurry Up" Yost, Michigan football coach. Admission
was ten cents for youngsters, forty cents for adults. At the Ford City
Theater, the feature was *Hell Bent for Heaven*, starring Patsy Ruth
Miller. Admission was ten cents for kids, thirty cents for adults.

There was no lack of meetings, either. At the Elks Hall at Fifth
and Main Streets, Taft Lodge No. 1527 of the Benevolent and Protective Order of Elks was holding its weekly meeting. At Masonic
Temple on North Street between Fifth and Sixth Streets, Midway
Lodge No. 426 of the Free & Accepted Masons was holding its
regular weekly meeting. At St. Mary's Hall at Third and Kern Streets,
Taft Council No. 2468 of the Knights of Columbus was holding its
weekly meeting. At the Boy Scout barracks at Fourth and Calvin
Streets, Taft Troop No. 1 of the Boy Scouts of America was holding
its regular weekly meeting, as was Cub Pack No. 2. At Smith Brothers Hall on Center Street, Taft Chapter No. 607 of Women of the
Mooseheart Legion was holding its regular semi-monthly meeting,
while in the Labor Temple Taft Carpenters Local No. 1774 was
holding its weekly meeting. Elsewhere in the Labor Temple, the
Federal Unions were holding their semi-monthly business meeting.

If anyone had suggested that the busy citizens of Taft might soon
tremble at the sight of mice, the person making the suggestion
would have been greeted with derision. The reaction, predictably,
when the first column of mice hit outlying oil camps in the Buena
Vista Hills between Taft and the lake bed was to regard the attack
as an uproarious joke. It was rumored on Center Street where the
principal business establishments were located that women were
frantically making plans to start a stilt factory.

The city of Taft in the heart of the Midway-Sunset oil field was a target for the advancing horde of mice in the winter of 1926-1927. (Photo from Kern County Museum)

On oil leases, the appearance of mice in outlying areas caused no immediate concern, nor did mice become any more important a topic of conversation than such things as the 16th anniversary sale at Smith Brothers' clothiers, during which A. B. Smith offered Blue Streak coveralls for only $1.39, workshirts for seventy-nine cents and canvas gloves for twenty-nine cents, or the preparations by Milham Exploration Company to drill another well at Buttonwillow. Less than a month before, Milham's Kern No. 1 had blown out. Gas had caught fire, destroyed the drilling rig and burned up three cars in which men in the crew had driven to work.

By early December, jokes about mice began to seem less amusing. Events took an ominous turn with accounts of mice invading beds and nibbling the hair of horrified sleepers, or chewing through the sides of wooden storehouses to devour food supplies, even crawling boldly into children's desks at Conley School. Apprehensive Taftians cast worried glances at the rodents' staging area, wondering when the attack would cease. From the lake bed, where 11,000 acres of barley and milo maize furnished plentiful fodder, a harvester sent back word that when he set his cutter low to harvest maize, he came upon mice so densely crowded that the blades "became choked with fur, flesh and blood to the resemblance of a sausage mill."

One of the first targets of the mice advance was Honolulu Consolidated Oil Company's main camp in the Buena Vista Hills, three miles from the lake bed. To turn back the onslaught, Superintendent Bob Maguire pulled oilworkers off usual jobs and assigned them the task of plowing furrows around derricks and other installations, then seeding the trenches with poisoned barley. Men with wheelbarrows went out each morning to carry away dead mice. It was estimated in one day alone as many as 50,000 mice were killed on the Honolulu lease. Still the mice kept coming.

"Fabled Pied Piper Needed," a headline announced in the *Driller* on Saturday, December 4. "Page the Pied Pier,"the newspaper article said. "The services of the fabled Pied Piper of Hamelin may be required to exterminate the hordes of rats and mice that are swarming over the city of Taft and surrounding leases.

"The visitation of the rodents has almost reached the proportions of a plague, and serious attempts are being made on some of the leases to exterminate them with poisoned grain and the use of huge traps.

"Housewives in the lower end of the city are complaining of the number of mice in and around their homes, although the greater number have swarmed to the leases outside the city.

"A theory has been advanced that the horde of mice comes from Buena Vista Lake bed and other lowlands where the water has risen on account of recent rains. This theory is supported by the fact that the leases lying in that direction were visited first by the army of mice and that they have now reached the lower end of the city. Taft Heights and South Taft are comparatively free of them to date.

"In the meantime there is considerable demand for mice traps, cheese and pussy cats."

Cats quickly proved ineffective. Once gorged, the felines showed little interest in rodents. One disgusted homeowner discovered sixteen mice sleeping under two cats.

Invading mice furnished the main subject for discussion three days later at the regular Tuesday luncheon meeting of the West Side Business Men's Club. On hand to inform those present of the mice situation were C. H. Bowen, Kern County's deputy horticultural commissioner, and Lloyd M. Sands, also a deputy. When someone suggested that the peak of the invasion might have passed, Bowen told the group the cold snap that had set in after recent rains might have slowed down the mice advance, but that the rodents would be back, "worse than ever." Bowen advised, "You'll need poison, and plenty of it." A. M. Keene, publisher of the *Driller*, moved that the club donate $50 for poisoned grain, and the motion was quickly passed. Clarence Williams, Taft's mayor, pledged the city's support in the battle against mice.

From another source came an offer of help which was gratefully accepted. The Automobile Club of Southern California offered its offices at 427 North Street as a command post through which poisoned grain might be distributed to rid the city of the invading army. Phil Lay, local manager of the club, announced someone would be on hand every day to dispense grain as required with the only charge to be that made by the county, which had agreed to supply poisoned grain at actual cost. Office personnel invoiced 600 pounds of poison in one-pound packages.

Meanwhile C. H. Bowen, with C. R. Hamilton of the auto club, toured the oil fields, calling on oil company superintendents in the

Oil installations in the Buena Vista Hills felt the first impact of the millions of mice moving out of Buena Vista Lake bed. (Photo from Kern County Museum)

Invading mice stripped the bark from small willow trees in the path of advance. See photo at left. Photo to the right shows milo maize stubble on the dry bed of Buena Vista Lake which helped sustain the army of mice. (Photos from Kern County Agricultural Commission)

affected area to marshal forces to combat the ever-increasing hoard of mice. Bowen reported that all concerned were interested in making a concerted effort to rid the community of rodents.

Oil companies turned to the task of blunting the mice attack with all-out determination. At the point of Buena Vista Hills, Honolulu Oil assigned crews to cut furrows extending for more than four miles. Whit Barber, Kern County's horticultural commissioner, led a platoon into the trenches to sow strychnined wheat. Standard Oil Company joined the fray, plowing a furrow one and one-half miles long to protect its installations. Midway Oil & Gas Company assigned a ditching machine to cut a circular furrow around its pumping plant. In a three-day period, the circular furrow accounted for more than 75,000 mice, among them, according to an account in the *Bakersfield Californian*, "genuine rats wearing shaggy winter coats." Shortly before Christmas, the mice attack abated. Weary men left trenches to spend Christmas at home.

Though the defenders had won a battle, they had not yet won the war. Even as residents of Taft and outlying oil leases happily returned to a semblance of normalcy, the mice regrouped. In Taft, plans proceeded for the Petroleum Club's golf tournament on the Sunday before Christmas. The winners would receive turkeys for Christmas dinners. For others who might want something besides turkey for Christmas, the California Market on Center Street announced a sale on buffalo meat from the Antelope Island herd in

Utah's Great Salt Lake. Two days before Christmas gentle rain fell, turning to snow that left a light mantle of white on the hills. Snow was followed by drizzling rain and temperatures as low as forty degrees.

Christmas passed, and New Year's followed with big dances at the Petroleum Club, Elks Club and Buchanan's Pavilion, where the management made good on promises to distribute "great loads of noisemakers, paper hats, serpentine and confetti to add to the pleasure of the evening and to assist in ushering in the New Year." The Shamrock Cafe offered a special New Year's dinner for $1.50, including California oyster cocktail, branch celery, ripe olives, chicken gumbo and a choice of chicken fricassee, roast domestic goose, roast duckling or roast young tom turkey, along with candied sweet potatoes, peas and carrots, strawberry parfait or hot mince pie, and coffee.

Early in January, mice in even greater numbers swarmed out of the lake bed, aiming massive thrusts not only at Taft but also at Maricopa, Elk Hills and its oil camps, the community of Tupman to the north, and at Paloma Ranch and newly-seeded farm fields to the east. Bolstering the attackers were millions of meadow mice, a hardier specimen than the house mice that had carried the attack before. Among newcomers were numbers of exceptionally large individuals, a not uncommon situation after periods of inordinate increase.

Advancing to the southwest, mice killed a penned sheep at Santiago Canyon near San Emidio Ranch, eight miles from the lake bed. B. M. Tibbett of Taft reported that a small army of mice were still eating the animal's flesh when he came upon the scene. Another column slipped past poison-filled trenches to touch off an exodus of women from Ford City, an unincorporated community adjoining Taft. Another column captured the Petroleum Club golf course, swarming over dirt fairways and oil sand greens with only token opposition from disgusted golfers. To the north, hordes advanced over the Taft-Bakersfield Highway. Thousands were ground to death under car wheels, making the highway dangerously slippery.

E. Raymond Hall, zoologist at the University of California, made a two-day survey of the mice's staging area and declared the rodent attack was "the strangest occurrence of its kind in the history of the United States." Hall did not allay apprehension on the West Side when he pointed out that according to mathematical calculations, one pair of mice could be responsible for the production of 16,146 others in a year's time.

Estimates of the rodent army placed size between 30 and 100 million mice, indicating attacking mice outnumbered residents of Taft by a conservative margin of at least 6,000 to one.

Followers of an evangelist predicted that the mice onslaught was the forerunner of the end of the world.

"Nation, State and County Battle Mouse Invasion," the *Bakersfield Californian* said in a headline on the January 15, 1927 edition. The newspaper reported, "Scene of what is declared to be the greatest rodent invasion in the history of the United States, the West Side of Kern County is today a trench-marked battlefield on which the U.S. Biological Survey, the State Department of Agriculture and county officials are fighting an advancing army of millions of field mice, meadow mice and ordinary house mice.

"Driven from the dry bed of Lake Buena Vista by recent rains, the rodent hordes have besieged Taft, Ford City, Tupman and the oil leases of Elk Hills and Buena Vista Hills.

"Close-packed battalions of the scampering mouse army have created the phenomenon of a moving landscape, observers report. But the mice move only to front line trenches, nine miles in length, dug in front of the advance by puffing tractors dragging deep bladed plows.

"Head over heels the rodents are tumbling into the trenches to feast on tons of poisoned wheat scattered by county authorities, dying before they can scramble to the trench tops.

"Approximately 100 mice to the foot is the toll that the death trenches have taken, according to L. M. Sands, county horticultural inspector, who has visited the battlefield.

"Whit C. Barber, Kern County horticultural commissioner, and C. H. Bowen, deputy horticultural commissioner, are in Taft today to direct the fight, and are being joined by rodent experts of the U.S. Biological Survey and State Department of Agriculture, specially commissioned to assist in the warfare.

"'There are lots of mice,' admits Mr. Barber, 'but the situation is well in hand, and the rodents should be exterminated within two weeks.'

"Highways in the foothills are dotted with thousands of bodies of mice that, though they escaped the poisoned grain, fell victims to automobile wheels.

"While the war is being waged in silence, squadrons of buzzards sail and swoop above the hills, hour after hour, lazily dropping to

From base camp on Pelican Island the "Mouse Marines" carried the battle to the rodents' lake bed redoubt. (Photo from Kern County Agricultural Commission)

Long thin line of "Mouse Marines" moved out to stop invading mice. (Photo from Kern County Agricultural Commission)

feast on the slain. Gorged, the birds are robbing the filling trenches and heaping the bodies in piles about the base of fence posts, supplying themselves for future feasts.

"One line of trenches was covered over today, it is reported, burying thousands of dead mice, and a new one hastily dug to check a fresh advance.

"Before the alarm became general, the mouse legions infested the oil leases near Taft and on the Honolulu Consolidated Oil Company property devoured quantities of flour, sugar and other supplies in storerooms."

The gray horde of mice raised the spectre of plague. Dr. William Stowe Fowler of Bakersfield offered reassurance. "Unless the mice are infested with fleas, which carry Bubonic plague, there is no danger of an epidemic," Dr. Fowler said. "I have lived in this county 27 years and was county health officer for six years, but I have never heard of a Bubonic plague death in Kern County. There was no trace of the plague in this county even when San Francisco was making strenuous efforts to abolish a spread of Bubonic, with Dr. Blue leading the plague fighters."

Two days later, the *Californian* offered readers far less reassuring news. "Bakersfield Menaced By Mice," the newspaper's headline said. "In close-packed herds, in tiny bunches, in straggling columns, singly and in pairs, Kern County's mouse army is marching toward Bakersfield," the paper reported. "The vanguard had reached Bellevue Ranch this morning, two miles west of Stockdale Ranch and just seven miles from this city."

After stating that the mice invaders had as yet done no great damage and that the situation at Taft appeared to be in hand, the article said that mice had swarmed onto Buena Vista Ranch, ten miles north of Buena Vista Lake, and had besieged Canfield Ranch after moving east along the Taft-Bakersfield Highway.

"Using inadequate mouse traps and all the poisoned grain they can get their hands on, farmers between Bakersfield and the stampeding mice are making every attempt to stop their advance," the article continued.

"Scurrying from their burrows in the dry bed of Lake Buena Vista, the mice have fled from the lake bed, apparently to escape floods from heavy rains, and traveled the following distances in their migration to higher lands: Lake west to Taft, eight miles; to Maricopa, 11 miles; north to Buena Vista Ranch, 10 miles, and northeast to

Bellevue Ranch, 15 miles.

"As they advance west, north and east, the mice are leaving behind thousands of their dead companions, stricken and bloated by poisoned grain placed in tempting piles and drifts in their path.

"Sunday the West Side highways were thronged with automobiles loaded with residents from all parts of the county, who traveled the battlefields of the mouse war to get a close glimpse of the attack of the gray horde and its subsequent destruction.

"Dell Holsom, superintendent of the Stockdale Ranch of the Kern County Land Company, declares that at Bellevue and Canfield ranch buildings are being protected by batteries of mouse traps, which are set in the evening and emptied the next morning. This morning every trap contained a mouse, and as many as 40 were captured in the vicinity of single buildings. But whether the traps and the sprinkling of poisoned grain will stop the march of the mice is a question in the minds of ranchers, it is said. So sudden was the advance upon Bellevue and Canfield Ranches that the poisoners were unable to dig a defense line of trenches."

While defenders feverishly cut new trenches and the Kern County Horticultural Commission stepped up distribution of poisoned grain to 1,500 pounds a day, word of the mice war spread throughout the United States. Offers of armies of alley cats poured in from all over the country. Matthew McCurry of the Society for Prevention of Cruelty to Animals in San Francisco offered a word of advice with regard to cats. "It's all right," McCurry said, "if you use country-bred cats, but city cats would be worse than useless because they don't know how to forage. They don't understand mousing. I speak, of course, of the majority of them. But country-bred cats would be fine."

One astute observer, aware that strychnine acted as a temporary stimulant before it killed, warned that six grains of poisoned wheat would cause a mouse to challenge a cat.

Many offered advice on how best to turn back the mice attack. The Army Chief of Chemical Warfare suggested use of poison chlorine gas. An Orange County, California woman recommended that vats filled with water and lye be placed in the path of the mice. After the mice had swum through, she said, they would lick their feet and die. The measure, she added, had been used with success in a rat plague in China. A Rushville, Missouri man suggested establishment of a state colony of skunks. Skunks, he said, would soon clear out the mice—and oil field towns too, added an oilworker. An Oakland woman advised that homeowners sprinkle essence of peppermint over carpets and in ice boxes because mice could not stand it and would leave. Frank Smith, Kern County clerk, received a letter suggesting that control workers soak bits of sponge in "sweet stuff," the theory being that the sponges would swell in the stomachs of the mice and kill them. One man wrote to the Chamber of Commerce suggesting that they shock the mice. "Why not run copper wires just above the ground and then turn loose high tension stuff through your wires," the man wrote. An electrical engineer figured the plan would cost from $12,000 to $15,000, a sum far in excess of any damage done so far. James Ogden, city manager of Bakersfield, received in the mail an enormous mouse trap sent by a Northern California friend.

Among those offering explanations for the movement of the mice from the lake bed was Rowen Irwin, a Bakersfield attorney. "Their god tells them it is going to be a wet year," Irwin said, "and they are fleeing for their lives from the lowlands of Lake Buena Vista into the hills around Taft. And there will be rain, and lots of rain, heaviest on the West Side."

Irwin said his prediction followed an internationally aired theory regarding the "intuition" of animals. "Look at Santa Barbara," Irwin said. "The morning of her earthquake, before the first tremor even, not a single bird could be found in her trees. All small animals had fled, it seemed, and the fact was noted by residents.

"On ships at sea rats have been noted to scurry up from below decks and leap into the water hours before trouble was discovered which threatened the lives of passengers.

"And in Lake Buena Vista lake bed the mice know that they will be overtaken by disaster unless they move, and move in a hurry.

"From the size of the migration it appears that the county is due to have one of the wettest years in her recent history."

Meanwhile in more than nine miles of trenches, a "they-shall-not-pass" determination among defenders, plus seven tons of poisoned grain, inflicted an estimated five million casualties on mice, halting the offensive at some points as far as ten miles from where it had begun. Though the advance was halted, there were still problems. One was seeding of crops. East of the lake bed, a farmer seeded ninety-one sacks of wheat in one day. The next day, every grain of wheat had been eaten by the mice. Federal and state officials were called in the next day and twenty-five sacks of wheat were seeded on the surface and plowed under. When officials and the farmer went to the field that night with lanterns, they found mice running around the field as thick as mosquitoes. By morning, the mice were gone and so was the grain.

A federal poisoner from the U.S. Biological Survey arrived from Denver, Colorado on January 22, 1927, to take command. His name was Piper—Stanley E. Piper—and he was a tall, serious man who took immediate offense at being hailed by newspapers as the Pied Piper. It was a joke he had undoubtedly heard before, for he had successfully turned back a smaller mice migration at Lovelock, Nevada in 1907-08. Piper promptly set up a base camp on Pelican Island, a low mound, in the northern portion of the dry lake bed, outfitted the camp with living quarters and cookhouse, and re-cruited a force of twenty-five men, promptly dubbed the Mouse Marines, to carry the battle to the rodents' redoubt. Assisting Piper was F. E. Garlough, from Berkeley, who was charged with making special studies of the best means of poisoning the rodents.

One facet of the scientific approach involved a census to deter-mine the magnitude of the problem. On one acre in the lake bed, a crew of men dug up all mice burrows or warrens and counted the number of mice. The official tally was almost 4,000. Projected to the 11,000 acres that had been planted to milo maize and barley and subsequently had spawned the mice horde, the figure indicated the presence of perhaps 44 million mice. In another test, one-third of an acre was dug up with a view of learning the percentage of the three kinds of mice found at the lake bed. On this test, 391 house mice, 119 meadow mice and 22 white-footed mice were uncovered, with ap-proximately twenty-five per cent estimated as still being under the ground.

Fleetest of the scurrying rodents were the house mice. Garlough's studies indicated that within forty-eight hours of the start of the mice migration house mice had showed up several miles away in such numbers that "housewives were gathering them in bucketfuls." Meadow mice spread more slowly, Garlough concluded, covering perhaps half the distance that house mice covered. As a result of studies, poisoners settled on steam-rolled barley as bait, seeding it throughout the area.

Nature, as if belatedly mindful of an obligation to preserve its balances, provided unexpected help. More than 1,000 ring-billed gulls appeared, diving out of the sky to destroy mice. Straggling companies of short-eared owls flocked into the mice redoubt, making nights at the Pelican Island base musical with their clear calls. Ravens, an estimated 400 of them, and hawks joined the airborne attack, among the latter such species as the marsh, American roughlet, western red-tail, Swainson, turkey vulture, Cooper, prairie falcon and sharp-shinned hawk. Other birds participating in lesser numbers included great blue herons, roadrunners, white-rumped shrikes and at least two golden eagles.

As the tide turned against the mice, envoys from the scientific community hurried in to study firsthand the phenomenal mice migration. Reports of the infestation carried by the press prompted the Rockefeller Foundation for Scientific Research to send a famed research specialist to the scene. The specialist was Clara J. Lynch, formerly of Johns Hopkins and more recently with the Rockefeller Foundation. One of the early finds was the identification of what was described as a highly unusual type of rodent among the migrating mice, one said to have "a blunt nose, blue-black rough hair, a small tail and small ears."

Beset by man and birds, mice fell back to short, blind excavations little resembling their normal systems of runways and tunnels. Advancing exterminators found evidence of cannibalism in the mice army. Epidemic disease, believed caused by bacillus of mouse septicemia, suddenly spread through rodent ranks. The great mice war ended in mid-February with losses calculated at more than 30 million mice and between $10,000 and $15,000 in damaged crops and expended poisoned grain.

In the spring, in fulfillment of Irwin's prophecy, nature provided a possible answer to why the mice had left their burrows. Heavy rains caused a larger than normal run-off from Kern River, flooding the once-dry bed of Buena Vista Lake.

Breaking the Pick

Even as Mr. and Mrs. J. N. Ripple busied themselves in preparation for Christmas of 1928 by stringing lights over the bare branches of a huge fig tree near their home on the Mascot lease on 25 Hill in back of Taft, Standard Oil Company's rigbuilders at a nearby location on the same lease went about the task of erecting a steel derrick that dwarfed wooden derricks over shallow wells drilled in years past.

Christmas was a week away, and like the year that was ending, it promised to be a good one, with stores like Asher's and Smith Brothers, the Ladies Toggery and Taft Furniture & Hardware in downtown Taft bulging with merchandise that would find its way under Christmas trees on the West Side. For friends back East, California Market offered the ideal present: a two-dollar gift box of San Emidio Champion oranges, grown only eighteen miles from Taft and famous the world over, featured recently on a broadcast over Salt Lake City's powerful radio station KSL that had been heard by a man named Pete Stromberg who had promptly cabled a request that oranges be shipped to him aboard the M/S *Stroch.* The ship was frozen in the Arctic's Amundsen Gulf at longitude 124W latitude 70N, 600 miles east of Point Barrow, Alaska.

To the string of lights on the fig tree, the Ripples added festoons of red and green crepe paper, tinsel and artificial icicles. To complete the display, Ripple spread lime on the ground beneath the tree and covered it with artificial snow to simulate the snow that normally could not be expected in Taft. The tree was the first lighted outdoor Christmas tree in the oil fields that holiday season. Since it was located atop the hill and at an intersection of three oiled roads, it drew its share of visitors, who marveled at the sparkling display in the clear December nights.

For Standard Oil, the steel derrick taking shape on the Mascot property was not so much an expression of the season, though there was stark beauty in the steel frame, as it was an expression of faith in the future, and an attempt to answer a question that had nagged oil

men ever since the first oil had been discovered on the West Side
more than thirty years before. The question concerned how deep
one could drill and reasonably expect to find oil. The question was
heightened by the persistent hunch that if a man drilled deep
enough, at depth he might find a bonanza that would surpass anything anyone had seen before. There were doubters, of course, but
there was growing evidence that the thought of finding oil below a
depth of one mile or more was not so far-fetched an idea after all.
For one thing, down in Texas a company named Texas Oil & Land
had drilled a hole that was down to 8,516 feet and according to
reports out of Big Lake was flowing 60 barrels a day of high gravity
oil. The hole was the deepest in the world, lacking hardly more than
2,000 feet of being two miles deep, and it was lending support to the
belief that the earth might hold oil at great depths.

If the thought of deep pay intrigued West Siders, no particular
area of the Midway-Sunset field had quite the same appeal as 25
Hill, where Standard now prepared to drill. Shallow wells there
were good ones, and because of the hill's prominent rise in back of
the city of Taft, it was constantly before men as a possible deep
prospect. Standard had every intention of finding what lay at depth,
and to that end the company recently had leased deeper rights from
Mascot Oil Company, the owner of the parcel on Sec. 26, 32S-23E.
Now, late in 1928, Standard was moving to drill the well that would
give up the answers.

Interest in the Mascot deep test was enhanced by the fact that
some preliminary moves had already been made toward seeing
what lay below proved sands, and there had been encouragement.
One of the first to drill for deeper sands was Victor Oil, which had a
parcel on the same section. Victor had gotten encouragement at
2,500 feet, but had not been able to produce the well, though the
show was good enough to boost the company's stock from sixty
cents a share to a high of two dollars. There had been trouble with
the water string. In cementing off, the bottom sand was plugged.
The company had come back up the hole to complete in shallow
sand. Even as Victor debated the pros and cons of drilling another
test, two major companies were rumored to be dickering with the
company for deep rights.

In the wake of the Victor test, another company named Italo
Petroleum had leased deeper rights on a 40-acre parcel on the same
section owned by Mt. Diablo Oil & Mining Company and was

Shallow wells on the Mascot property on 25 Hill were good producers. Many thought the real bonanza lay deeper. (Photo from Clarence Williams Collection)

drilling ahead enroute to a proposed depth of 3,000 feet. Italo had had shows just below 1,900 feet that were good enough to bring a stream of customers to Ed Lawton's window in the local branch of the Bank of America with many orders for stock purchases. The rush was big enough to cause another flurry in Victor stock. The market was bullish, not only in California but across the nation. It seemed obvious that the sky would be the limit if a discovery were made. The work of erecting the steel derrick ended, and three days after Christmas Standard began rigging rotary tools.

On New Year's Eve, the West Side celebrated the beginning of a new year that promised to be even better than the one ending with a public dance at Buchanan's Pavilion and private parties at the Petroleum Club, Elks and Sciots. Five days later, company crews spudded in to drill the Mascot No. 1, accompanied by the rumor that the big rig was going deeper than other tests on 25 Hill, that it would in fact go deeper than any well drilled before in the Midway-Sunset field.

A few years before, any talk of drilling a hole much deeper than one mile would have evoked disbelief. As late as 1925, a prominent consulting geologist from New York City, a Dr. Leith, had presided over a round table at which participants including various oil company geologists had concluded that there would shortly occur an acute shortage in the world's oil supply.

When one prominent outsider sounded a more optimistic note, he had been met with some ridicule. The outsider was Thomas Edison, who had said that the oil industry would presently drill a mile and a half or more to find immense new reservoirs. "Why they're not doing that already," the electrical genius had said, "is difficult to understand."

An editorial writer on *Petroleum World*, a California oil publication with offices in Los Angeles, quickly took Edison to task. "Oil men are not anxious to increase the balance on the debit side of the ledger by drilling one and one-half miles for oil that they have no means of knowing is in the ground at that depth. An invitation should be extended Mr. Edison to visit the next annual meeting of the American Petroleum Institute. Even oil men have been known to learn a great deal about the oil business from these meetings."

As if an impending shortage of oil were not bad enough, the industry also faced serious technological problems. One was inability to go to great depths to search for oil. It had been an astonishing event when an operator in the Athens area of the Rosecrans field at Los Angeles succeeded in drilling to a total depth of 7,591 feet. It was even more astonishing when the well flowed 200 barrels a day, a world's record for deep production. The discovery of oil at that depth made it seem that oil men might be missing a bet by not going deeper.

The prospect of going deeper was an awesome one. "It is not particularly attractive," said K. C. Heald, a well-known oil man, "but it must be accepted with good grace." Heald described California and the Gulf Coast as "the great bottomless areas in the United States" and said oil was not to be hoped for in beds more than 300 feet below the Ashton zone at Huntington Beach or more than 2,500 feet below the Brown zone in the Long Beach field, nor more than 3,750 feet below the proved sand in the Inglewood field, all in Southern California.

"Only the Southern states, Western Oklahoma and Kansas, parts of the Rocky Mountain states and California can reasonably be considered fair hunting grounds for the man who is willing to drill 6,000 feet or more," Heald said. "And even in these states, the facts now available strongly indicate, if they do not absolutely prove, that the deepest possible producing sands will not be more than 7,000 feet below the surface where they cap anticlines, although on the flanks of anticlines, these sands may be at greater depth." And what

In 1930, Taft celebrated its twentieth birthday with a gala parade and a world's depth record at Standard Oil Company's Mascot No. 1 on 25 Hill. Driving a Thomas Flyer in the parade was Glen Alexander. In the back seat with feet propped on a box of Giant Powder Company explosives are Blanche and Nellie Alexander. (Photo from Clarence Williams Collection)

if an operator dared to drill deeper? "Operators who drill where the rocks are too much altered will be puzzled and disgusted by finding heavy asphaltic material in place of the highly fluid oil they are looking for."

All those gloomy predictions had been in the recent past. As the bright new year began, all things seemed possible. Only a few months before, Charles A. Lindbergh had flown across the Atlantic Ocean. Technology was moving ahead in fields other than aviation, too. Actors in movies had begun to talk. Four days before Standard spudded in to drill the Mascot well, Taft's Hippodrome Theater had shown its first talking picture. The movie was Warner Brothers' *State Street Sallie*, starring Conrad Nagel and Myrna Loy, and while only the more dramatic scenes were sound sequences, the first all-talking picture, *The Singing Fool*, starring Al Jolson, was coming to the Hippodrome on January 14, 1929 for a three-day run, with two shows each day, including a 2:30 P.M. matinee and a 6:30 P.M. evening show.

Nationally, the country was prosperous. The prospect facing the president-elect, Herbert Hoover, who had carried the West Side by a better than two-to-one margin over his Democratic opponent, New York's Governor Al Smith, seemed to be for more of the same. At the moment, Hoover was off on the battleship *Maryland* on a goodwill cruise to South American waters. He had recently enjoyed a friendly visit with the chief of state of Brazil which included a tour of Rio de Janeiro.

As one index of prosperity, the Automobile Club of Southern California reported on the phenomenon of the two-car family. More than three million American families, the Auto Club said, now owned two cars, and the number was steadily growing. The club's spokesman said there were various reasons for the growth of the two-car family, including such things as membership in golf clubs, children at college, the increasing number of women drivers, and the use of automobiles in business. The matter of cars might not have been the soundest indicator elsewhere, but in Taft it was important, for people there prided themselves on the cars they drove. The Buick agency in Taft as the new year began reported sales were up: eight cars in recent days, including a sedan to Standard Oil for company use.

If national prospects were bright, equally bright were the prospects for California oil. Only two months before, Milham Exploration Company had been drilling a wildcat on the Elliott lease at Kettleman Hills, seventy-five miles northwest of Taft, when at a depth of 7,108 feet the well blew in, sending a plume of oil and gas into the sky to signal the discovery of what appeared to be an enormous new field. Two Taftians had journeyed to Kettleman and come back to report on what they had seen at a Wednesday luncheon of the Taft Rotary Club at the Shamrock Cafe. Fred Doyle and Richard Shinn told Rotarians they thought fifteen wells would be drilling at Kettleman within the next few months and that it was the belief of oil men that the new field might prove to be of the size of Midway-Sunset. They described the awe-inspiring sight of the Milham well, which was still flowing an estimated 4,000 barrels a day of clean 60-gravity oil, and told of plans in the area to build a road from Hanford to the new field.

From the *Daily Midway Driller*, January
1929. (Beale Memorial Library)

From Santa Fe Springs in Southern California, the news was
equally exciting. Only months before operators had begun to go
deeper, and for their efforts they had completed twenty-three
deeper zone wells, developing 75,000 barrels a day new production.
At the moment, twenty rigs were going up in the field, and something like 215 wells were reported to be drilling for the deep sand.
With Santa Fe Springs as an example, property owners in the nearby Huntington Beach field were mounting an effort to raise $60,000
for a bonus to be given the first wildcatter to find deeper sands in
their field.

If deeper production could bring a boom at Santa Fe Springs and
offer the possibility of one at Huntington Beach, there seemed no
reason why it could not do the same for Midway-Sunset. Because of
developments in the Los Angeles Basin, the West Side field had
been relegated to third place among California's fields after a reign
of more than ten years as the most productive field. Thanks to
Signal Hill, Long Beach was now the top field, putting out 190,000
b/d, followed by Santa Fe Springs, 92,500 b/d, and Midway-Sunset,

75,000 b/d. Deep oil might be the answer to putting the West Side back on top. Standard was not the only company with sights set on deeper pay. Others were taking leases for deeper oil from Taft to Fellows, among them Union Oil Company, Barnsdall Oil, Associated and Shell. There seemed no mistaking the fact that Midway-Sunset was going to get a run for its money.

At Mascot No. 1, Standard Oil Company quickly left no doubt it intended to do a first-rate job. After making 150 feet in hard sand and boulders, the crew went in to take the first core, cutting five feet, recovering a disappointing two inches of hard blue shale with fine lenses of oil silt. They went in again with the core barrel, coring a 20-foot interval, recovering six feet of gravel with slight oil saturation and odor throughout. Back in with the drilling bit, they made 25 feet, then went in with the core barrel, coring the next 400 feet of hole, recovering 362 feet, most of it a matrix of pebbles, sand and clay with slight oil saturation, grading to sandy green clay and fine tar sand. At 608 feet, the company halted downward progress to ream. Thirteen days after spudding in, they cemented an 18⅝-inch water string at 565 feet. After the Division of Oil & Gas approved the water shut-off, they went back in to resume coring, alternately drilling and coring to a depth of 997 feet, at which point they cemented 13⅜-inch casing with 280 sacks, the last 250 treated with calcium chloride.

Emphasis on coring underscored Standard's determination to take no chance of walking away from pay sand. It also meant that a clearer picture of the Midway-Sunset field would be developed. Coring was a relatively new technique in the oil fields, having come into its own less than ten years before in development of Southern California fields, where Elliott Core Drilling Company, headed by a former Shell Oil Company geologist named John E. (Brick) Elliott, had made its first successful coring run in the Huntington Beach field in 1921. The technique had been used with marked success in the Santa Fe Springs field, and Standard was anxious to see what it could do at Midway-Sunset.

After cementing the second string of casing, Standard's crews went back to drilling and coring ahead, reaching a depth of 2,794 feet by February 4, one month after spudding in, at which depth they lost circulation, that is, lost the drilling fluid circulated into the hole to bring cuttings to the surface, coat the walls of the hole, counteract subsurface pressures and lubricate the rotating bit. To regain circulation, crews labored to thicken the drilling mud with Aqua Gel, a mixture of silica and aluminum which when mixed with water formed a viscous or jelly-like substance. They also added sawdust to the mud stream, using 11,300 pounds of it before they finally succeeded in stopping the loss of drilling fluid. Coring ahead, they recovered 57 feet of shale with fractures carrying pyrite, better known as fool's gold, and the odor of sulphur. Trouble had just begun.

Four days after the first bout with lost circulation, they were battling the same problem again, solving it this time with light Muroc mud and Aqua Gel. They cored ahead to 3,129 feet and lost circulation again. With Aqua Gel and clay-like Mojave mud, which was heavier than the Muroc mud, they pumped in six bales of hay, 258 sacks of sawdust and more than one ton of cottonseed hulls. The battle was won in two days, and they warily drilled and cored ahead, recovering hard brown sand and shale.

On March 22 while drilling at 4,110 feet, they lost circulation again. Out came the trucks, carrying the sacks of material drilling crews would have to mix with the mud stream. Over a four-day period, tired crews labored to pump in 156 tons of dry Mojave mud mixed to a weight of approximately 80 pounds per cubic foot when co-mingled with 800 barrels of mixed mud, 290 sacks of cottonseed hulls, 17 bales of straw, 410 sacks of sawdust and 35,400 pounds of Aqua Gel. They regained circulation several times, but each time lost it again. Five days after running into trouble, they pumped down a Perkins cement plug with an eight-foot wood guide on bottom. The plug stopped at 1,173 feet, and they dumped in three sacks of cement treated with calcium chloride. After the cement hardened, they cleaned out to 1,173 feet. The hole stood half an hour without mud loss, and they cleaned out and began drilling and coring ahead. There was another bout with lost circulation at 1,359 feet, another cement plug set, more cleaning out and renewal of drilling and coring from 1,200 feet, making new hole below the plug.

All told, circulation was lost twelve times more before they reached a depth of 6,355 feet, at which point the hole was in hard brown shale. On August 25, 1929, they cemented 9-inch casing at 6,350 feet, using a plain shoe, that is, a reinforcing collar of steel screwed onto the bottom joint of casing to prevent abrasion or distortion of the casing as it forced its way past obstructions on the wall of the hole, with Baker-Burch cement float guide.

An early January 1930 storm dropped a record four inches of snow on Taft, threatening to shut down drilling operations at the Mascot well, leaving a blanket of snow on the town, including Taft Union High School. (Photo by Maurice Bejach)

After allowing the cement to stand for a week, crews cleaned out plugs and cement to 6,330 feet and bailed fluid to 3,000 feet. The well stood six hours without a rise. They cleaned out to 6,357 feet and bailed to 2,953 feet. The well stood for twenty hours; fluid entry was equivalent to half a barrel a day. The Division of Oil & Gas approved the water shut-off, and crews resumed drilling and coring ahead. With pipe in the hole, lost circulation in the fracture zone would no longer be a problem. Crews pursued the search, making hole through hard brown shale, picking up the first gas and oil showing on the mud ditch just below 6,600 feet. The show consisted of bubbles of gas and a stain of oil in the mud stream circulating out of the hole.

While the cementing of casing solved the circulation problem at Standard's Mascot well, there was another problem facing not only Standard but other producers as well, and it would not prove that simple a matter to solve.

The problem involved growing oversupply of oil. While production surged, demand failed to keep up until the amount of oil available far exceeded the amount the market could absorb. There was talk of cutting back, of shutting in wells and curtailing production, and everyone agreed it would be a good thing. The disagreement came on whose production was to be shut in.

When the suggestion was made that production from Kern County fields be cut back by some 4,000 barrels a day, no one argued that it would not be a wise move, but the question remained, whose production should be cut? West Side producers had a good candidate. They suggested production be trimmed in the Kern River field on the East Side of the county. Kern River operators complained that they already had several hundred shut-in wells and were plainly doing their part. They countered with a suggestion that production at Midway-Sunset be cut fourteen percent.

In Southern California, 300 operators met and announced plans to cut back production at Signal Hill by 41,000 b/d and at Santa Fe Springs by 50,000 b/d. Whose production was to be cut was not made plain, and no volunteers stepped forward. The same dilemma plagued operators in other oil-producing states. Several weeks later when the American Petroleum Institute announced production figures for the United States, it was noted that production had risen to 2,760,000 b/d, up 400,000 b/d in a year's time.

In May, President Hoover called an oil conference for the following month. The conference was to be held at Colorado Springs, and those invited were the governors of the western producing states, or their representatives, whom it was hoped would lay the groundwork for creation of an interstate compact to better control drilling and eliminate waste.

Early in June, governors and their representatives met in Colorado Springs with Secretary of the Interior Dr. Ray Lyman Wilbur representing President Hoover.

Whatever hope there was of working out an early end to the over-supply situation fell short when those at the conference split into two factions. One faction representing independent producers held out for the inclusion in any agreement that cut production of a clause providing for a tariff provision against the import of foreign oil. The meeting broke up four days after it began without agreement on a course of action.

Gasoline wars erupted across the country. In Los Angeles, the price dropped from 16½ ¢ per gallon to as low as 10½ ¢. The American Petroleum Institute attempted to make points out of the situation, noting from its New York offices that gasoline prices were at their lowest level in eleven years, averaging 17.52¢ per gallon compared with an average of 22.63¢ per gallon in the preceding years. The API pointed out that gasoline was unique among commodities in that its price had declined despite enormous gains in demand. With the price of gasoline clearly a bargain, Standard Oil Company announced in San Francisco that in the interest of conservation it was cutting the price it would pay for crude oil at Santa Fe Springs, Signal Hill and Seal Beach by 50¢ to 75¢ a barrel.

Though oil was in oversupply, nationally the economy seemed in solid shape, with more news emphasis on the problems of enforcing Prohibition and the gang wars attendant on it, including a Valentine's day massacre in Chicago in which seven men were killed, than with economic ills. There were a few disquieting signs, like the closing of ten banks in Florida by the state examiner and the failure of three banks in New Jersey to open their doors following an investigation by the state's Department of Banking & Insurance.

On the last Friday in October, the Associated Press sent out a story that Washington was keeping an eye on Wall Street, but though securities prices had slumped, the unofficial opinion was that the drop need have no depressing effect upon the nation's general business structure. In Taft's *Daily Midway Driller*, the Associated Press story was buried under a headline that dealt with that night's football game between Taft High's Wildcats and Fresno High's Warriors.

The following Tuesday, desperate speculators sold 16,400,000 shares of stock.

A day later, Associated Press reported that the stock market was returning to normal with scores of stocks on the New York Stock Exchange up $5 to nearly $30 a share. In Washington, D.C., R. Julius Klein, assistant secretary of commerce, went on the radio to say that American business need expect no adverse results from the preceding day's collapse of stock prices. Klein said that less than one percent of the population had been affected by what he described as the "speculation gyrations."

On the following day, K. R. Kingsbury, president of Standard Oil Company, issued a reassuring statement. "I know of nothing in the present condition or the future of the oil industry that justifies the extraordinary depreciation in prices of oil stocks today," he said. "No new conditions have arisen threatening our company or, so far as I know, the industry. On the contrary, the prospect for effective conservation, which will result in balancing production with demand, is brighter today than it has been at any time."

Six days later, stocks dropped again, declining $5 to $30 a share in an abbreviated three-hour session on the New York Stock Exchange.

Another six days later, prices on leading issues went down one dollar to twelve dollars a share. More than 100 stocks sold at new low prices for the year.

As selling continued, President Hoover announced a plan to seek stabilization by tapping the $250 million fund established by the Jones-White Act to assist in expansion of the merchant marine. Shipyards would be booming in six months, newspapers reported. A week later, representatives of the nation's public utilities reported plans to spend $1.5 billion in improvement and expansion of their systems during the coming year. A few days later, Ford Motor Company announced a wage raise totalling $20 million, boosting the minimum wage for its employees from $6 to $7 a day. The sell-off continued. By the end of the year, the stock market crash had cost investors an estimated $40 billion.

Through the months that misfortune was befalling the east's financial establishment, there seemed no immediate reason to believe that there would be a depression in the oil fields. On 25 Hill, Standard's crews continued drilling and coring ahead, making steady footage through hard shale and sands, getting some shows though the deep bonanza they sought seemed always to elude them.

The West Side Businessmen's Club met at the Shamrock Cafe, and the *Daily Midway Driller* reported that a spirit of optimism prevailed. Taft merchants professed to see a period of real prosperity ahead. Some mention was made of the fact, as the newspaper had pointed out not long before, that Taft might well be the richest city per capita in the United States with a $2 million payroll each month. Deposits in the town's three banks totalled $6,716,000—or $2,024.72 for each person counted in the last census. There were 4,830 cars registered, or one and one-half cars for each man, woman and child.

In mid-October 1929, the service clubs of the West Side sponsored a banquet at the Woman's Improvement Club as part of a nationwide "Light Golden Jubilee" program to honor Thomas Edison. Two hundred and fifty West Siders turned out to hear Dr. Willis White of Bakersfield speak on the life of Edison, who had invented the incandescent electric lamp fifty years before. Even as the community honored the aging electrical genius who had wondered why the oil industry did not drill deeper wells, Standard's Mascot was below 7,000 feet and making hole.

On the business scene, an exciting new enterprise came to Taft: regularly scheduled air service. Continental Air Express of Los Angeles inaugurated service with a Lockheed Monoplane that could carry five passengers. The plane left Los Angeles at 8:30 A.M., arriving in Taft at 9:40 A.M., departing at 5:00 P.M. for the return to Los Angeles. In effect, the schedule allowed for an almost full day of business in the oil fields. One-way fare was $9.90 for the hour's flight. Two months after service began, the airline put a tri-motored Fokker on the run, changing the schedule to make Taft the intermediate stop on morning and afternoon flights between Los Angeles and San Francisco. The 62-foot-long Fokker, weighing 9,173 pounds fully loaded, could carry eight passengers.

Drilling passed the 7,800-foot mark at the Mascot well, where Standard crews ground away at hard silty sands and hard shales, coring almost as much as they drilled. Geologists checked cores, noting some light oil and gas shows in shale fractures in the Devilwater formation. By Christmas, the well was below 8,000 feet.

From Southern California came word of a new world's depth record. Oscar Howard's Hathaway No. 7 near the corner of Telegraph and Norwalk Roads at Santa Fe Springs had gone to 9,350 feet and was testing.

The news was overshadowed by an oil field disaster. At Standard's No. 61 on Sec. 7, 32S-24E, one mile northeast of Taft, a crew was pulling tubing when an explosion occurred, engulfing the rig in flames. Two men died almost immediately. Two others died within a day of burns. Only one member of the crew escaped. The tragedy added the names of C. S. Grady, Pat Harmon, Arthur Nance and Edward Wathen to the list of sixty-three men killed that year in oil field accidents in California. Grady left a wife, two sons and a daughter; Harmon, a wife and two sons; Nance, a wife, two daughters and a son; and Wathen, a wife and son.

Three days later, the newspaper carried a short notice of another economic disaster in the east. Stutz Motor Company, whose products long had been admired in the oil fields, was going into bankruptcy.

Hardly had the new year of 1930 arrived than nature provided an unaccustomed spectacle. A storm dropped a record four inches of

Fast-flowing wells like the one this crew has just brought in spelled trouble for California oil as supply overwhelmed demand. Crew took time for picture after laying down traveling block, foreground, as part of rigging down. (Photo from Clarence Williams Collection)

snow on Taft, threatening for a time to shut down drilling operations at the Mascot well. At the rig, men bundled up and talked of snows they had seen in other places. It was the greatest snowstorm in Taft's history. The job went on.

Though Standard continued to make hole, cutting through hard shale that carried a faint odor of light oil, there were increasing signs that the oil fields might not be immune to the economic problems afflicting the rest of the country. There was a hollow ring to pronouncements of good times coming, of bank statements showing fine gains, of new markets opening for California oil in Japan, when such statements were measured against a sharp drop in building permits, refineries going on a six-day week to hold down products, and talk of a six-hour day on drilling rigs. The six-hour day, proponents said, meant four crews a day would be required for round-the-clock drilling operation, keeping a maximum number of men employed, even if each man earned less. For the operator, the arrangement was said to offer one notable advantage: men would not have to be given time off to eat as on eight-hour tours.

In March, Standard passed 9,200 feet, coring continuously as crews had done since 8,251 feet, making about fifteen feet a day, requiring twelve hours or more for each round trip to pull the core from the hole and go back in to cut another.

Late in the month, the crew cut a core taking the well to 9,629 feet, making Standard's Mascot No. 1 the deepest well in the world by almost 300 feet. Only one foot was recovered from the 18-foot interval cored. A. D. Henderson, Jr., described the core as "very hard poorly sorted medium grained gray sand, faint odor light oil, no cut with carbon tetrachloride, fair cut with Acetone."

Of the depth record, *California Oil World* noted, "Geologic data, some of which may prove of the utmost scientific value, constitutes practically the only value of the deep drilling Standard is carrying on in the Mascot well." The depth was so great, the writer continued, "even if oil were found now, it would require a big well with ample gas pressure to flow it to make it of commercial value." However, the writer was not entirely pessimistic. "This well being the first to tap the earth to this depth, everything that its log reveals will be of importance. The Mascot may develop information of great value not only to the geologist and to the scientific world but also to the driller and to the manufacturer of drilling equipment." The writer concluded, "When oil is higher in price, even as deep a well as this might pay for operation."

In April, crews ran the longest string of casing ever run, cementing 5¾-inch casing to a depth of 9,302 feet. They installed a Shaffer-Warner casing head, closing in the well.

Standard Oil Company announced that it was suspending the operation until July 1931, more than a year away, or later, in compliance with a plan by the company and other majors to hold down activity in California fields "to eliminate the over-production which has featured the industry for some months." Officials were reticent to give out details of shows the deep well was rumored to have encountered.

In the same issue of the *Driller* that reported the shutting in of the deepest well in the world, there was a report that unemployment in the oil fields was small with only 163 men out of work out of a census population of 3,096 persons. At the time, the town was within two months of having its first marathon dance at Buchanan's Pavilion. Seven couples would compete for $1,000 in prizes in an attempt to break the world's marathon dance record of 1,685 hours. None would break the record, though two couples would succeed in dancing forty minutes out of each hour, resting twenty minutes, for more than 280 successive hours, or almost twelve straight days.

Before the year ended, more than six million Americans were out of work. The number rose to twelve million during the following year. While Mascot No. 1 stood shut in, the country sank steadily into the worst depression in its history. Millions of persons lost every cent they owned. Banks failed, factories shut down, stores closed, and almost every business seemed paralyzed. More than 5,000 banks failed, and over 32,000 businesses went bankrupt. Farm prices fell lower than ever before. Desperate men sold apples on street corners, ate in soup kitchens and lived in collections of shacks like the one of cardboard, corrugated iron and scrap lumber that took shape at the east end of Center Street in Taft and was called Hoover City.

It was not until the summer of 1932, more than two years after the Mascot well had been idled, that Standard Oil began to rig rotary again at the steel derrick on 25 Hill. Though there were still hard times ahead, production in California had been cut to 460,000 b/d, some 225,000 b/d less than it had been when the crash occurred, and it was time to go back to work. Standard and others began calling back the drilling crews who had struggled to survive with whatever jobs they could get through the long night of the depression.

While the Mascot well had been shut in, the world's depth record had passed out of the country, belonging now to Penn-Mex Fuel Company's Jardin No. 35 in the state of Vera Cruz, Mexico, which had gone to 10,585 feet.

On July 6, 1932, Standard's crews began the long road back at the Mascot No. 1. Running in with a fish tail bit, they were able to get no deeper than 9,190 feet at which point drill pipe would not rotate, apparently because of crooked hole and heavy mud.

Intending to clean out to 9,288 feet to make a casing test, crews cleaned out in error to 9,306 feet, bailed and found a 60-foot rise of gassy foam. They ran two-inch tubing and swabbed, getting gas-cut fluid. Swabbing to 5,700 feet, they recovered some oil that tested at 29 degrees gravity. They swabbed to 7,200 feet without getting commercial entry. They filled the hole with fresh water and shut it in, leaving it idle through most of August.

In September, operations resumed. Crews ran in with 2⅞-inch drill pipe to make more hole. Alternately drilling and coring, they went to 9,685 feet, twisted off and recovered drill pipe. They cored a two-foot interval from 9,734 to 9,736 feet and recovered four inches that F. W. Ohliger described as "hard black shale stratified with many thin bands gray limey shale. Black shale contains microscopic globules asphaltum."

Driller M. D. Freeland at the controls of Standard Oil Company's Mascot No. 1.
(Photo from *Petroleum World*)

Standard Oil Company's Mascot No. 1, the world's deepest well. (Photo from *Petroleum World*)

On October 22, 1932, they had gone to 9,753 feet and were trying to get back on bottom with a 4¾-inch Reed bit when they were blocked by a bridge of protectors. Efforts to clean out the bridge failed. They could get pipe in the hole, but could work up only inches at a time and then with maximum allowable strain and great difficulty. They could go no deeper. With difficulty, they cleaned out to 9,710 feet and ran in with a Johnston formation tester to test on bottom. A small amount of gas flowed. They recovered thin drilling fluid at a rate of forty barrels a day.

Laboriously, crews continued to try to test, encountering bridge plugs of rubber protectors, getting formation entry that plugged the bit while making connections, opening it by reversing circulation.

Before November ended, they had bailed to 8,250 feet, recovering brackish water with only traces of black oil. The die was cast. In December, crews rigged up cable tools. Continuing to test, getting no encouragement, they slowly came back up the hole. In March 1933, they began cutting and pulling casing to salvage what they could from the unsuccessful well.

Rotary drillers who had been assigned to the well during the long drilling job included C. A. Ohler, H. M. Carroll, C. T. Wachob, J. L. Dykes, Albert Larson, Walter Goff, Sam Landrum, T. E. Dorsey, H. P. Snyder, Grover Hixon, M. D. Freeland, C. E. Black and L. J. Frey.

Standard—or cable tool—drillers included R. W. Patterson, J. F. Richardson, Thomas L. Ray, H. Archibald, O. M. Hixon, Albert Larson, B. F. Magee, C. E. Elder, C. A. Ohler, W. B. Gordon and J. W. Tomer.

By June 9, 1933, abandonment was complete. A month later, R. I. Brown, district petroleum engineer, signed the well log, and Mascot No. 1, once the deepest well in the world, passed into history, taking with it the days when all things had seemed possible.

Index

232

233